D0172365

DATE DUE

NATIONAL GEOGRAPHIC
BIRDING ESSENTIALS

Siskiyou County Library
719 Fourth Street
Yreka, CA 96097

NATIONAL GEOGRAPHIC
BIRDING
ESSENTIALS

ALL THE TOOLS, TECHNIQUES, AND TIPS YOU NEED
TO BEGIN AND BECOME A BETTER BIRDER

JONATHAN ALDERFER AND JON L. DUNN

NATIONAL GEOGRAPHIC
WASHINGTON, D.C.

CONTENTS

Adult male **Painted Bunting**, opposite *(Texas, April)*; alternate adult white-morph **Reddish Egret**, previous page *(Florida, April)*

CHAPTER 1

THE PLEASURES
OF BIRDING

We all seek out nature at some point in our lives, whether for peace and contemplation or just out of basic curiosity. As our world becomes more crowded and the pace of life quickens, many of us experience a strong desire to connect with the natural world. Being in the presence of a tree, a flower, a butterfly, or a wild animal can satisfy an elemental need, and our lives feel more complete.

Of all the animals in nature, birds are the most visible, and they are among the most favored of Earth's creatures to study and enjoy. Recent studies tell us that just in North America, millions of people, perhaps even tens of millions, enjoy watching birds. Of the 10,000 or so species of birds currently recognized in the world, slightly under 10 percent, or about 935 species, have been found in North America, north of Mexico. They coexist with us in a bountiful variety of shapes and colors—from

A **Great Egret** graces the marsh at Anahuac National Wildlife Refuge in Texas *(opposite)*. Observing birds connects us with the wider—and wilder—natural world we live in.

common garden birds to the rare and exotic—and animate the landscape we live in.

What is it about birding that generates so much interest? Many of us are drawn to birding out of an appreciation of the beauty of birds and their variety of songs. The sight of a dozen or more species of warblers on a spring migration day, all with their own spectacular colors and distinctive patterns, is unforgettable. The long migratory journeys that certain species undertake—many thousands of miles, for some—and the mystery of how they reach their breeding and wintering grounds with such precision is awe-inspiring. There is also the excitement of the "treasure hunt": What is out there today? What will I find? Every outing offers the chance of finding and identifying a rare bird, a species very unusual for your area or even for North America as a whole. For those of us who have lived through a long winter with few birds among the bare branches, there is a thrill of anticipation

This is as simple as bird identification gets: a male **Northern Cardinal**, bright red with a black face, a thick red bill, and a prominent crest. After a short time birding, you'll get to know the easy-to-identify birds in your area and not even think about their field marks. They'll be recognized instantly, like an old friend, and new or unusual birds—even if you can't identify them—will stand out more vividly. (Texas, November)

when we see the greening of the forest, the budding of vegetation, and the arrival of the first spring migrants. We know that within a week, or several weeks, the trees will be full of birds and birdsong. The renewal of the forest to color, activity, and song nurtures our own annual renewal.

The simple act of looking for and identifying birds will sharpen your perceptions and intensify your entire outdoor experience. While birding, you'll discover that you see everything more acutely—the terrain, the season, the weather, the plant life, other animal life. Most of us enjoy the challenges of the identification process and putting names to the birds we see. Some birds (the male Northern Cardinal, for example) are easy to identify. Others species, such as dowitchers, are much more difficult to identify. A few species remain nearly impossible to separate in the field with 100 percent certainty, even by experts. One of the aims of this book is to help the beginning birder get acquainted with the basics of bird identification—what to look for and how to use the clues you see to identify the birds you find.

As you become more involved in birding, you'll discover that you are part of a large fellowship of birders, many of whom enjoy participating in organized field trips and gatherings of birding clubs and organizations. It is a community of people bound by a love of nature and of birds in particular. Most birders have diverse interests and enjoy other types of nature study as well. For instance, when bird activity slows down at midday, there may be butterflies and dragonflies to see and identify. And evening doesn't bring an end to the possibilities: Spotting owls, mammals, and various snakes is often best done in the evening hours. Come to think of it, there is really no need for the complete birder to sleep at all!

Birding doesn't require long sojourns into wilderness areas. It can be enjoyed virtually anywhere—a city park, a local marsh, or even your own backyard. For instance, New York City's 843-acre Central Park in urban Manhattan is one of the most renowned birding spots in North America during migration.

Installing bird feeders near your home is a sure way to get some species to come to you. Feeders, which come in a variety of styles and sizes, provide close-up views and a great introduction to the birds that share your immediate outdoor space. Many birders enjoy keeping a yard list—the number of species you have seen in or from your yard. In fact, during the past ten years, one yard in Cape May, New Jersey, has recorded well over 300 species of birds.

Once you've started down the birding path and proceeded past your backyard and local parks, whole new horizons will open up to you. Those little maps in your field guide make it abundantly clear that an Elegant Trogon is not going to visit your backyard in Massachusetts, nor will a Sooty Shearwater be coming anywhere close to your apartment in Denver. You will realize that to experience these species, and hundreds more, you're going to have to bird new places, both near and far. Visiting special places with special birds—such as southeast Arizona (home of that Elegant Trogon), the lower Rio Grande Valley of Texas, the Florida Keys, and California's Salton Sea—is one of the joys of going birding. If you need one, here is the perfect excuse to visit some of the wildest and most scenic areas in North America—or the world.

For beginners, birding is easy, relatively inexpensive, and full of discovery. All you need is a midpriced pair of binoculars, a field guide, and a desire to experience and learn about the birds around you. The adventure of birding can last a lifetime.

The Conservation Community

Putting a name to a bird is the first step in preserving and protecting it. Without names, birds are generic and often ignored, but once

Here's a challenging identification: a mixed flock of worn, breeding-plumaged adult dowitchers in late summer. There are two species in this photograph. Compare the single **Long-billed Dowitcher** *(lower right)* to the two *hendersoni* **Short-billed Dowitchers** *(center)*. The plumage of the Long-billed shows bars rather than spots on the sides of the breast and is a little more richly and extensively colored below; structurally, it has longer legs. *(New Jersey, August)*

A male **Elegant Trogon** hides most of his crimson breast but reveals another, more subtle side of his gorgeous plumage. This much sought-after species is uncommon and localized in pine-oak woodlands of southeastern Arizona—one reason a trip to the area is on every birder's wish list. *(Arizona, May)*

you attach a name to a species, both it and you are transformed. For then you can consider this particular bird's nesting requirements, its feeding niche, its migratory pathways, and its singularity; and you care about its welfare. Because of our connection with birds, it's natural for many of us to become active conservationists. As a birder, your grassroots awareness of the local environment will place you among those best informed about conservation issues in your community, whatever your political affiliation.

You may want to get involved with local, national, or even worldwide conservation organizations. At the local and state levels there are many opportunities to become involved, from serving on public planning commissions, to joining the local chapter of the Audubon Society, to forming your own group or club. In addition, there are numerous national and international organizations that focus primarily on birds. Created to fund research, habitat conservation, and education, they are worthy of your consideration.

Even if you choose not to join an organized group, you should be aware of your economic clout as a member of the unofficial fraternity of birders. As a group, birders spend hundreds of millions of dollars every year just in the United States. When you travel, don't be bashful about letting people know that you are looking for birds. Those communities that host special birds and special habitats benefit economically from our visits. Good stewardship of natural areas does not have to run counter to local business interests. Wear your binoculars proudly when you travel.

How This Book Is Organized

This book is divided into nine chapters, each of which contains detailed information on the skills you'll need to become a birder or to improve your birding. For beginners, we have provided basic information on equipment, parts of a bird, how to make an identification, fieldcraft, and other topics. We hope that more experienced birders will find in these pages something new or perhaps a new take on something old.

We always start a chapter with an overview of the basic information before we tackle more complex issues. The photographs in this book have been chosen to illustrate specific points, and in most cases the captions accompanying the photographs present detailed information that expands on the text.

First and foremost, birding is a visual activity. Our aim, therefore, has been to make this book as visually rich as possible, and to achieve this we have chosen images from some of the continent's best bird photographers. We hope you find yourself returning to the photographs and their captions both to learn from them and to enjoy them.

This pair of endangered **Whooping Cranes** consists of a juvenile (left) and an adult, mostly likely a young bird with its parent. Whooping Cranes nest in remote freshwater marshes of Wood Buffalo National Park in Canada and migrate as family groups to the marshes on the Texas Gulf Coast, where they spend winter. This population is now about 230, and slowly increasing. These magnificent cranes draw birders from around the globe to the Texas coast. (Texas, November)

CHAPTER 2

GETTING STARTED

I n this chapter we offer some suggestions for developing your birding skills. You don't need to pursue all of these ideas; be guided by
what you enjoy doing most. Later in the chapter we'll provide advice
on both choosing and using the basic equipment that all birders
need to have: binoculars and a field guide. At the end of this chapter,
there are sample pages from a field guide with all the features annotated.

**White-breasted
Nuthatch** *(above)* is a
common resident through-
out much of North
America, but its calls and
plumage vary regionally.
Work to determine
whether more than one
species should be
recognized is ongoing.
Birders *(opposite)* take in
the avian sights at High
Island, Texas, a premier
location for migrating
songbirds, especially
in spring. *(above:
Connecticut, March;
opposite: Texas, April)*

The Learning Process

How you develop as a birder is a personal quest. For every
birder, though, three areas of experience form the matrix of
the essential store of knowledge: looking at birds in the field,
learning from others, and studying books and other references. How quickly you acquire your store of knowledge is
up to you; you can devote all your free time to birding pur-
suits or advance at a casual pace.

Look at Birds Everyone has seen or heard birds. But not
many people look at birds closely or even notice them regularly. Even fewer go searching for them or try to put a name to them.
That's what birders do. Just get outdoors, find a bird, hoist your binoculars, focus on it, and you've become a birder with field experience.
This bird has entered your memory bank of images, even if you don't
know its name. Over time, depending on how often you go out looking, those mental images become linked with a bird's species name,
even its age, its sex, and other details of its life history. The effect is
cumulative: The more you go birding, the faster you'll progress.

Expert birders often identify birds without a field guide or even a
good look; they have a highly detailed memory of many species built
up over years of field experience. Some small, partially seen plumage
detail or behavior may be all that's necessary for them to make an identification, especially with birds they expect to see in a particular area.
This ability can only be gained by looking at birds in the field; you

cannot get it from a field guide, and it takes time to acquire. Fortunately, time spent in the field looking at birds is also a lot of fun.

Go Birding with Other Birders Many birders start off birding alone, relying on their own nascent abilities to puzzle out bird identification. Accompanied only by a field guide, they are likely to find the process slow going. Compare the following two scenarios.

You're on your own and see a small, brown-streaked bird in an overgrown field. It's not familiar to you, but it's about the size of a sparrow, so you decide to start with sparrows in your field guide. You find about 30 species with the word *sparrow* in their names and still more that look like sparrows but are called something else. Since you're in Connecticut and it's July, the range maps help you narrow the possibilities down, but only a little. By the time you're ready for another look—to check for head stripes, wing bars, or breast streaking—the bird is gone. It's frustrating. If you keep at it and see similar birds a number of times, you'll start to build a mental image of this bird and remember details of its plumage. Eventually you figure out that you are seeing Grasshopper Sparrows—a hard-won, but still tentative, identification.

Here is another scenario: Your local chapter of the Audubon Society offers a half-day field trip for beginners, and a local, long-time birder will lead the trip. Over the course of the field trip you get to know the others in the group and share some of your experiences and growing enthusiasm for birding. The leader takes you to great birding spots that you didn't know about (and that you'll want to come back to on your own) and helps everyone get a handle on some tricky identifications. When that small, brown-streaked bird shows up, the leader identifies it as a Grasshopper Sparrow and tells the group that they nest in this field. He suggests that you look for its flat head and large bill; he goes on to point out its streaked head with a pale median stripe, its unstreaked breast, and its short, pointy tail feathers. When the bird disappears, the leader describes its buzzy, insectlike song, which helps you get another look; later you compare your observations to the text and illustrations in your field guide.

To jump-start your identification skills, that last scenario works better than the first. You'll often go birding alone, and you will probably see many birds that you are able to identify. But a group field trip or an outing with a birding friend allows you to check on your identification skills. Working out an identification with another birder will help both of you. Simply talking with someone about a bird that both of you are watching will help focus your attention, and you're less likely to give up on a challenging identification. Even when alone it can be

helpful to describe in a low voice the birds you observe. Just pretend you're talking into a tape recorder.

Fieldcraft—how to move, where to look, how to use your equipment—is another skill best learned by birding alongside someone experienced *(see chapter 8)*. Feel free to ask fieldcraft questions on a field trip during a break in the active birding. Most veteran birders are delighted to share their knowledge.

Some new birders prefer to learn on their own, at their own pace, or just don't like groups; some experienced birders feel the same. If you are a new birder and only bird alone, you may not progress as quickly, but if you keep at it you'll learn. Later, you may have to unlearn habits or misinformation that might have been avoided had you gotten feedback. But, on the other hand, you'll make your own discoveries.

Study Your Field Guide at Home When you are out birding, concentrate on the birds you see, not on the illustrations and text in your field guide. Experience the living, breathing bird in front of you. Look at its behavior, study its shape, describe its patterns and plumage details to yourself; keep looking as long as it remains nearby. If you have a really cooperative bird, do it all again or look at specific areas of the bird. Maybe you'll even have time to write some notes or make a sketch.

We recommend an even more radical approach: Leave your field guide at home or in your car, where it won't distract you or interrupt your experience. Not having it with you will force you to look critically and

Grasshopper Sparrow belongs to a group of grassland sparrows in the genus *Ammodramus* that are notoriously difficult to observe. They are easiest to see when the males are singing on the breeding grounds. Grasshopper Sparrow's songs are high, buzzy, and insectlike. *(Connecticut, July)*

to work at remembering what you've seen. The end result is that you'll learn faster. When you return home, check your field guide against your observations. Try to do it while the experience is still fresh.

That said, your field guide is your primary reference and an essential piece of birding equipment. To use it effectively, you need to become intimately familiar with how it is organized *(see pages 26–27)*. Browsing your guide at random is fun and productive. If you keep an extra copy next to your favorite chair, you'll pick it up at odd moments and dip into it. Soon you'll flip right to the thrushes or the terns, remembering that the towhees are with the sparrows and the mergansers are near the end of the section on ducks. You'll find that what once seemed random has an orderly (and scientific) pattern.

Tufted Puffins are spectacular, large-billed seabirds that nest colonially on islands and coastal cliffs in the North Pacific—from northern California to Alaska and west to Japan. Puffins are related to gulls and terns and can be found after them in your field guide. *(Alaska, July)*

Bird Your Yard Wherever your yard is or whatever the view from your window, you'll see birds, even in urban areas. During spring and fall migration, the variety can multiply. Most birders enjoy keeping a yard list, a record of what they've seen in and from their yards; many also record the dates when migratory species return in spring or leave in fall. Your yard can be a private proving ground for your growing skills. Here are simple ways to attract even more birds to your yard.

■ **Feeding Birds** Many people start birding when they put out a bird feeder and then try to figure out the identities of the birds that show up. Birdfeeding is extremely popular in North America, and why not? For a minimal investment of time and money, you can have close-up views of some fantastic birds. Most of them will be common residents, but the occasional "good bird" (an unusual one) is always a possibility. Choosing from among feeder types and foods is not hard, and many how-to books cover the subject well. Local chapters of the Audubon Society often sell birdseed and feeders and give out good advice for free; dollars spent there will support local conservation efforts. Start out by offering black-oil sunflower seed in a hanging feeder; add a thistle (niger) feeder for small finches, a platform feeder for ground-foraging birds, and a sugar-water feeder for hummingbirds. Before long, the birds in your area will be regular visitors.

■ **Water** A reliable water source is very important to birds. Birdbaths with fresh water will attract many species, including some that won't come to feeders. Install a simple dripper above your birdbath—a suspended gallon milk jug with a pinhole in the bottom will drip for hours; the sound and movement of the water will attract even more species. During migration, this may attract birds like warblers and vireos. More elaborate setups that work automatically but require an electrical connection are available, as are devices to keep birdbaths free of ice.

■ **Landscaping** Enhancing the plantings in your yard to attract birds is more involved, but it can yield impressive results. If you are able to, consider planting fruit trees, vines with berries, and flowering shrubs. Ask your local nursery to recommend native plants. While you're at it, ask about plants that attract butterflies. If you shrink the area you devote to lawn, you'll create a more natural environment around you.

Keep Notes on What You See Most birders enjoy keeping track of their birding experiences. Usually this involves keeping a list or multiple lists—life lists, year lists, yard lists, state lists, county lists, and so on—that form a personal history of your birding experiences. Your lists bring back fond memories and remind you of what you still want to seek out. And by keeping a journal of your sightings, the process becomes a true learning experience.

A tube feeder loaded with black-oil sunflower seed will attract a wide variety of species. This one is being visited by a **Carolina Chickadee** (top), two **American Goldfinches** (center), and a male **Northern Cardinal** (bottom). (Ohio, January)

We encourage new birders to keep notes right from the start. (In chapter 8, we discuss how to keep a journal and sketch a bird.) The simplest form of note taking—which is also more informative than keeping a checklist—is writing a list of species with the number (or an estimate) of individuals seen at a specific location and the date. Add a line or two noting the time you spent, whom you were with, the weather, and the type of habitat, and you have a basic journal entry. Many times that is all you will need or want to do. Keep entries in a notebook and turn the page when you move to new location.

A great time to write a description (not just a number) is when you see a new species, a species you don't see very often, or a common species doing something unusual or in an unusual plumage. For those important sightings, try to note all the details you can. It's best to do this while you are looking at the bird or at least before you consult a field guide. Describe what you actually see, avoiding descriptions such as "it looked just like the illustration in the field guide." Sometimes you will only see enough to write a partial description. That's okay. Don't fill in the blanks by guessing or recalling the field guide illustration. Describing your impression of a bird's behavior or shape can be as important as describing the color or pattern of its plumage. These are field marks, too, even though describing them with exactitude can be difficult.

You can even practice describing birds at home: Look at a bird photograph in this book, cover the caption, and write down a description of what you see. Afterward, you can compare your description with the one in your field guide. You don't have to know the species' name before you write a description, but it will help if you know the names and locations of specific feather groups, such as primaries, secondary coverts, and tertials. These terms are important to a birder's vocabulary, and to communicate clearly, you need to be familiar with them. We have devoted a whole chapter to the subject *(see chapter 4)*. Adding a simple sketch to your description, with lines pointing out the bird's features, is very effective and less wordy.

You might think, *Why bother? What can I add to what is already known?* But that misses the point. While you are probably not going to discover a new field mark, you are training your eyes to see and documenting your firsthand experiences. If you've never taken notes, written a description, or sketched a bird, you'll be amazed at how the whole process sharpens your skills and how much more you retain when you put your observations on paper.

Binoculars

To see birds well, to identify them with any certainty, to simply enjoy them in all their stunning detail, you need good birding binoculars. Really good binoculars will last for decades and become a natural extension of your eyes and hands.

Here is a piece of advice for those new to birding: Those old binoculars that have been lying around the house for years will not do. Most will be of low quality, out of alignment, and unsuitable for birding. Directly compare them to entry-level birding binoculars (about $200), and you will be convinced. Birding doesn't require a big outlay of cash, but this is the place to spend what you can for quality. With good birding binoculars you will see more detail with less eyestrain and "get onto birds" faster. Many people struggle briefly with old or cheap (under $100) binoculars, and some give up birding without ever experiencing the pleasure of looking through good binoculars. Once you know that birding interests you, don't wait to upgrade.

Choosing Binoculars Most long-time birders have strong opinions about binoculars: which have the best optics, which are easiest to use, which are overpriced. The list goes on. A good approach to choosing binoculars is to become familiar with the features important to all birders and then make an informed decision within your budget. Definitely try out as many models as you can. Seek out other birders' advice; field trip participants are often delighted to let you try out their binoculars and tell you what they do and don't like about them. We will mention a few brands and models, but there are many good choices.

Below is a list of features to become familiar with. This is not a treatise on binoculars; our aim is to give you some basic information and get on with the birding.

■ **Magnification** All binoculars feature a set of numbers, such as 8x32 or 10x42. The first number is the magnification (or power) of the binoculars. The second number is the diameter in millimeters of the objective lens (the lens at the front). For instance, 8x binoculars will make the bird you are looking at appear eight times larger compared to your view with the naked eye. For birding, binoculars below 7x do not have enough magnification, and anything above 10x is too difficult for most people to hold steady. Of the few zoom models available (with a variable range of magnification), most work well only at the lowest power, and they are overpriced, heavy, and a poor choice.

7x binoculars With the best depth of field (so you don't have to adjust the focus wheel as much) and the largest field of view (so you see more area), these are great for finding and tracking birds in wooded areas. Many birders find them underpowered in some situations.

10x binoculars These provide the largest image of the bird you are looking at. You will see more detail if you can hold them steady enough, for they magnify any vibration or shake more than lower power binoculars. Their field of view is usually less expansive than that of lower power binoculars, so you might have more problems locating birds or tracking a nearby bird in motion. As a class, they do not focus as closely and have a shallower depth of field, so you have to adjust the

Roof prism binoculars may look sleek and simple on the outside, but this cutaway view reveals a complex system of lenses, prisms, and gears. The very best optics, like these **Zeiss FL**, come with a hefty price tag. Good entry-level binoculars start at about $200. *(courtesy of Zeiss Optics)*

Nikon Medallion 8x21

Nikon 7x35 Action BJ

Swift Ultralite 10x42

Pictured above are three sizes of **porro prism binoculars**: a compact reverse porro model, with the objective lenses closer together than the ocular lenses (top), mid-size (center), and full-size (bottom). Although not as durable as the roof prism design, porros are less expensive, and many models have excellent optics.

focus more often. They excel in open situations, such as looking at distant shorebirds or raptors.

8x and 8.5x binoculars This is the middle ground and the most popular choice of magnification among birders. The best models rival 7x models in close focus, depth of field, and field of view, and they have an extra bit of magnification.

■ **Construction** The two basic designs for binoculars are porro prism and roof prism. The internal arrangement of prisms accounts for their different shapes; excellent binoculars are manufactured in both styles. **Porro prism** These traditional-looking binoculars have wide-set objective lenses. Porros are less expensive to manufacture, so you can find excellent choices in the low- to mid-price range. The design produces a more three-dimensional image—an attractive, underappreciated feature. However, the design is inherently more fragile and more likely to go out of alignment, and some birders find them uncomfortable to hold. **Roof prism** This style has sleek, straight-through barrels, but on the inside they are actually more complicated. The roof prism design fits most people's hands better, feels more balanced in use (steadier binoculars mean better image and more detail), and is lighter in weight. Roofs are also more durable and easier to waterproof. The very best and highest priced binoculars employ this design. On the negative side, internally they have more glass surfaces, and lower priced roof prism models are often optically inferior to comparably priced porro prism models.

■ **Image Quality** This is hard to quantify and somewhat subjective. When you pick up binoculars you want the image to have a natural, restful quality and to be easy to adjust. Some binoculars have "jumpy" images or ones that feel like they are squirming when you pan across the landscape; this is caused by a lack of sharpness at the edge of the field of view. Other defects to avoid are a noticeable color cast to the image and color fringing (as evidenced by multicolored edges around objects). To assess image quality, compare similarly priced models side by side.

■ **Image Brightness** Image brightness is directly related to the size of the objective lens—the light-gathering lens at the front of the binoculars. Check the size of the objective lens by looking at the second number featured on all binoculars; for example, 10x42 binoculars have a 42 mm diameter objective lens. The size of the objective lens divided by the power, called the exit pupil (42 ÷ 10 = 4.2 mm), gives a rough estimate of image brightness. (Higher numbers are better.) Brightness is also affected by the quality of the glass and the coatings. (Fully multi-coated lenses are best.)

■ **Size and Weight** The size of the objective lens is also the variable that most affects both the overall size and the weight of binoculars.

For everyday birding, most birders prefer midsize (±32 mm objective lens) or full-size (±42 mm objective lens) binoculars. Some high quality, midsize binoculars are very close in image quality to comparable full-size ones, and the savings in weight can be substantial—around 20 percent—although the prices are about the same. If possible, handle models in both sizes and choose the model that best fits your hands. A midsize model will definitely feel better around your neck. Some very good compacts are available (8x20 and 10x25 are popular sizes), but few birders use them as everyday binoculars.

Leica Trinovid 8x20 BCA

■ **Field of View** A wider field of view is better. A wide field of view makes it easier to find birds and easier to keep moving birds in view. Most manufacturers label all their birding binoculars as wide-angle, but all are not created equal. The standard measure is how many feet from side to side you can see at a distance of 1,000 yards. Anything under a 330-foot field of view starts to feel narrow; a 400-foot field of view feels luxuriously expansive.

Vortex 8.5x32 Spitfire

■ **Close Focus** Closer is better. This feature has improved markedly in recent years; a few models allow you to focus on your feet! That might be more than you need, but the growing number of birders who also look at butterflies find this highly desirable. At the least, you don't want to find yourself backing away from a bird so that you can focus on it. That translates into a close focus distance of 10–12 feet; do not accept a distance greater than that. Less than 8 feet is excellent.

Swarovski 8.5x42 EL

■ **Fast Focus** Preferred by most birders, this feature is easy to check. Start with the binoculars focused as close as possible and then turn the focus wheel until you have focused on a distant object. If you had to turn the wheel much more than a full revolution, the focus is not fast. While you are doing this notice whether your finger falls naturally on the focus wheel; if it doesn't, try a different model. On some "fast focus" models, the focus wheel has been replaced by a rocker bar. These are useless for birding and impossible to focus accurately.

Three sizes of **roof prism binoculars** are shown above: compact *(top)*, mid-size *(center)*, and full-size *(bottom)*. The roof prism design fits most people's hands nicely, and is employed in all the very best and highest priced models.

■ **Use with Eyeglasses** If you wear eyeglasses, you need to be able to use your binoculars with them on. This may limit your choices. The specification you need to check is eye relief. Think of it as the distance (in millimeters) that you gain by folding down or sliding in the eyecups. Because your glasses hold the binoculars' ocular lens farther away from your eye, there needs to be a way to compensate for this; otherwise the image you see will appear severely narrowed, like you are looking through a tunnel. Most eyeglass wearers require about 16mm of eye relief, but check the image with your glasses on.

■ **Waterproofing** Waterproofing is an excellent, increasingly common feature, especially on roof prism models. Not only does this protect them from an accidental dunking, but problems with internal fogging are also alleviated. Remember that water resistant is not waterproof. Only waterproof binoculars can survive immersion in water.

■ **Armoring** Most birding binoculars now come with some type of rubber armoring. Armoring will protect the binoculars from minor dings and feels good to most birders. It will not protect them from a drop onto a hard surface; if that happens, immediately check the alignment.

If you wear glasses, make sure that the model of binoculars you purchase has enough eye relief for a full view. High price is no guarantee of good eye relief. Try the binoculars out before buying them.

■ **Price** Although birding binoculars range in price from around $100 up to $1,800, it is convenient to group them into three price ranges: $100–$500, $500–$1,000, and over $1,000. When you compare different models, compare those in the same price category. A $1,000 pair will almost always be noticeably better than a $200 pair—that's no surprise—but among the various $200 models there will be some that are clearly superior to others. Brunton, Bushnell, Leupold, Nikon, Pentax, and Swift are all respected manufacturers that offer binoculars in the low- and mid-price ranges. Nikon and Brunton also compete in the high-price range.

If you ask knowledgeable birders to recommend the very best binoculars, three models emerge: Zeiss FL, Swarovski EL, and Leica Ultravid. These high-end models are priced around $1,500 and up. There are minor differences in their specifications, but at this highest level of quality, comfort in your hands or even the appeal of the styling can be a deciding factor.

■ **Purchasing** Once you've determined the exact model you want, do some comparison pricing. Shopping by mail order or at an Internet site might save you money, but you won't be able to try out the exact pair you are buying. Wherever you shop, remember to ask if the binoculars have a full domestic warranty and to inquire about the return policy. A local chapter of the Audubon Society or another nearby nonprofit conservation organization also may sell the model you want.

Customizing Your Binoculars There are two adjustments that you should make to your binoculars so that they are customized to your face. Failure to make these adjustments will dramatically reduce the quality of the image you see, no matter how good your binoculars are.

■ **Adjusting Barrel Width** This adjustment is very quick and simple. You want to adjust how far apart the binoculars' two barrels are to match how far apart your eyes are (interpupillary distance). Simply squeeze the barrels closer together or move them farther apart until you see the largest image with the least amount of black around it. The optimum image area will be oval shaped. Binoculars should be able to hold that adjustment while you are birding, without your having to constantly fiddle with them. If the hinging action is too loose, the only solution is to have it professionally repaired.

■ **Adjusting the Right Eyepiece** This adjustment is a bit more complicated. You need to adjust the individual eyepieces, or oculars, to your eyes. The right-side ocular is the adjustable one. Turning the right ocular makes the adjustment (known as diopter control); on some models an adjustment wheel is located on the center hinge with the focus wheel.

Here's the procedure: While making the adjustment, keep both eyes open, and wear your glasses if you normally do. Cover your right eye and simply focus on an object in the middle distance (30–60 feet). Then cover your left eye and, without using the focus wheel, make the diopter adjustment to the right ocular until that same object is in sharp focus. There is usually a dial with tiny numbers on it that allows you to remember what your personal setting is, in case it gets moved.

Some binoculars with adjustments on the right ocular have trouble holding the setting: That is, the ocular turns and the setting shifts during birding use. You can stop this unwanted rotation with a small piece of tape, a heavy but small-diameter rubber band, or just the right size O-ring from a hardware store.

■ **Optional Adjustments** The straps your binoculars came with can be upgraded to a more comfortable, padded strap. Some birders swap out the straps for a harness that distributes the weight more evenly and stops the binoculars from swinging. If you keep the straps relatively short, they will bounce around less and you'll reduce by a microsecond the time it takes to lift them to your eyes, which can make a difference.

Your binoculars will function as a crude microscope if you use them in reverse: Hold one of the ocular lenses very close to what you're looking at and look through the large objective lens. Try it out on a feather and examine its structure, or check out small insects.

■ **Cleaning** Do not clean the lenses too aggressively; doing so can scratch the surface. At home, use canned air to blow off grit, a soft brush to loosen anything left behind, and another blast of air. Then fog the lenses with your breath and gently wipe in a circular motion with an optical cleaning cloth. In the field, blow off the grit as best you can, fog them, and use a piece of soft, 100 percent cotton cloth to wipe them.

Using Binoculars Everyone knows the basics of using binoculars, but a few birding techniques are important to know. With time, these and others you discover on your own will become almost second nature.

■ **Aiming** You will soon learn that you don't find birds with your binoculars; the field of view is just too small. You find birds with your naked eye and then aim your binoculars at them. This is sometimes called "getting onto a bird." Movement or a familiar silhouette may be your first clue that a bird is present. At that point, keep looking at the bird (don't look down for your binoculars; they're still around your neck) and lift your binoculars to your eyes. With practice, the bird will be right there or close by; the trick is to never stop looking at the bird. Sometimes it's easier to find an adjacent and slightly larger landmark—such as colorful vegetation or anything distinctive that you can remember for a moment—and use that to get onto the bird. Roll the focus wheel slowly back and forth; the shallow depth of field may have turned the bird into an out-of-focus blob. If you don't see the bird (keep looking for movement) in a few seconds, lower your binoculars, locate the bird again, and repeat the process.

Your binoculars' shallow depth of field can sometimes be useful. If you see movement but underbrush or leaves obscure your view, focus on the movement anyway. Then roll the focus wheel slowly back and forth, and the bird may pop into view. The obscuring vegetation is so out of focus that it melts away.

When you're out birding, try prefocusing your binoculars to the distance at which you are seeing the most birds—close by while walking down a brushy trail, middle distance in the woods, or infinity in open situations. This can save time, and you'll get good looks at more birds.

■ **Scanning** Sometimes scanning with your binoculars before you actually see a bird does work. During hawk watches, sea watches, and pelagic boat trips, there are often periods of time when nothing is in view. Try scanning far into the distance with your binoculars prefocused on infinity and slowly working the area to pick up birds invisible to the naked eye. If you're on land, sit down and rest your elbows on your knees. The steadier your binoculars, the smaller the speck you will be able to resolve. You probably have an idea about where birds are likely to come from, such as along a distant ridge at a hawk watch or near the horizon on a boat trip. With luck, distant specks will come nearer, but even if they don't, you might see enough for a probable identification. Experienced hawk watchers are adept at long-range identifications. Most raptors have distinctive silhouettes, flight styles, or other long-range field marks, and there are not many species to consider. Scanning the edges of ponds and other bodies of water can also pay off. Distant birds might be roosting or blend in so well that you can't see them with the naked eye. When you find something, maybe you can move closer or use a telescope.

Field Guides

You need at least one field guide. Birders rely on their collection of field guides to help them identify all the birds they see. As a consequence, field guides have evolved into hard-working, highly organized, information-delivery systems. To get the most out of them, you need to know how they are organized and what they can and cannot do.

A Very Brief History The early descriptions and paintings of North American birds from the 1700s and 1800s, including John James Audubon's masterful paintings and engravings, have little to do with today's field guides. The first step toward a usable North American field guide came in 1895, when ornithologist Frank Chapman published *Handbook of Birds of Eastern North America,* followed by *Color Key to North American Birds* in 1903 with drawings by Chester Reed. Reed went on to write and illustrate what many consider to be the first truly pocket-size field guide to birds in 1906, *The Chester Reed Guide to the Land Birds East of the Rockies.* Reed later produced more guides on a variety of subjects. Then in 1927 Ralph Hoffmann published his *Birds of the Pacific States* with illustrations by Allan Brooks. This small book described birds as if you were in the field looking at them—for the first time ever. Seven years later, in 1934, Roger Tory Peterson published *A Field Guide to the Birds,* and the evolution of the modern field guide took a giant step forward. The Peterson guides set the standard for many years. The popularity of birding grew rapidly in the latter part of the 20th century,

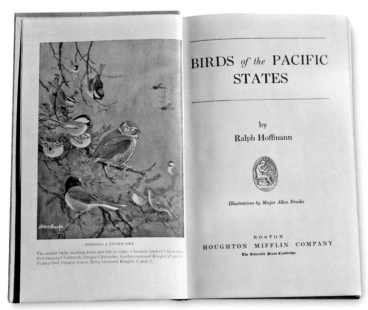

Ralph Hoffmann's *Birds of the Pacific States* (1927), illustrated by Allan Brooks, was one of the first field guides to describe how birds look in the field. It also included extensive behavioral information. In 1934 the first Peterson guide was published; today there are a great variety of field guides to choose from.

A Guide to the Guide

These two pages show a sample spread from the *National Geographic Field Guide to the Birds of North America,* fifth edition. The various features are annotated around the periphery. Many field guides have similar layouts.

Family Account: Very brief text on features shared by all species in the family, followed by the worldwide number of species in the family and the number of species found in North America.

Scientific Family Name

74

Grebes (Family Podicipedidae)

A worldwide family of aquatic diving birds. Lobed toes make them strong swimmers. Grebes are infrequently seen on land or in flight. *Species: 22 World, 7 N.A.*

Scientific Name (or Binomial): Genus and species together form a species' unique name.

Least Grebe *Tachybaptus dominicus* L 9¾" (25 cm)
A small, short-necked grebe with golden yellow eyes, a slim, dark bill, and purplish gray face and foreneck. **Breeding adult** has blackish crown, hindneck, throat, and back. **Winter** birds have white throat, paler bill, less black on crown. In flight, shows large white wing patch. **Range:** Rather uncommon and local; may hide in vegetation near shores of ponds, sloughs, ditches. May nest at any season on any quiet, inland water. Casual straggler to southern Arizona, southeast California, south Florida, and upper Texas coast.

Range Map: The different colors delineate breeding (red), winter (blue), and year-round (purple) ranges.

Pied-billed Grebe *Podilymbus podiceps* L 13½" (34 cm)
Breeding adult is brown overall, with black ring around stout, whitish bill; black chin and throat; pale belly. **Winter** birds lose bill ring; chin is white, throat tinged with pale rufous. **Juveniles** resemble winter adult but throat is much redder, eye ring absent, head and neck streaked with brown and white. First-winter birds lack streaking; throat is duller. A short-necked, big-headed, stocky grebe. In flight, shows almost no white on wing. **Range:** Nests around marshy ponds and sloughs; sometimes hides from intruders by sinking until only its head shows. Common but not gregarious. Winters on fresh or salt water. Casual to Alaska.

Length: Given in inches and centimeters (most useful for comparisons to other species). Wingspan is given for species often seen in flight.

Horned Grebe *Podiceps auritus* L 13½" (34 cm)
Breeding adult has chestnut foreneck, golden "horns." In **winter** plumage, white cheeks and throat contrast with dark crown and nape; some are dusky on lower foreneck. Black on nape narrows to a thin stripe. All birds show a pale spot in front of eye. In flight (next page), white secondaries show as patch on trailing edge of wing. Bill is short and straight, thicker than Eared Grebe's; neck is thicker too, crown flatter. Smaller size and shorter, dark bill most readily separate winter Horned from Red-necked Grebe (next page). **Range:** Breeds on lakes and ponds. Winters mostly on salt water but also on ice-free lakes of eastern North America; a few winter inland in West. Casual in Newfoundland.

Species Account: The text describes the important field marks and notes age-related and seasonal differences.

Eared Grebe *Podiceps nigricollis* L 12½" (32 cm)
Breeding adult has blackish neck, golden "ears" fanning out behind eye. In **winter** plumage, throat is variably dusky; cheek dark; whitish on chin extends up as a crescent behind eye; compare with Horned Grebe. Note also Eared Grebe's longer, thinner bill; thinner neck; more peaked crown. Lacks pale spot in front of eye. Generally rides higher in the water than Horned Grebe, exposing fluffy white undertail coverts. In flight, white secondaries show as white patch on trailing edge of wing. **Range:** Usually nests in large colonies on freshwater lakes. Rare in eastern North America.

Range Description: The text often includes a brief habitat description and mentions the abundance level.

English Name

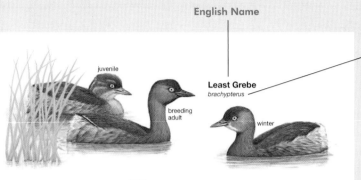

Least Grebe
brachypterus

juvenile

breeding
adult

winter

Subspecies Name:
Usually given if the
illustration depicts a
known subspecies or
subspecies group.

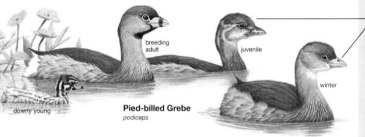

breeding
adult

juvenile

winter

downy young

Pied-billed Grebe
podiceps

Illustrations: All
markedly different
plumages are depicted.
For Pied-billed Grebe,
breeding adult, winter,
juvenile, and downy
young are shown.

Horned Grebe
cornutus

adult in
spring molt

darker
winter

breeding adult,
with "horns" raised

breeding
adult

winter

Horned   Eared

winter

Eared Grebe
californicus

1st fall

paler
winter

downy
young

breeding
adult

winter

Reduced-Scale Figures:
These show variation,
additional plumages,
specific behaviors, or
comparisons to other
species.

Main Figures: The main
figures on the same page
are all shown at the
same scale.

Male **American Goldfinch** *(left)* and male **Wilson's Warbler** *(right)* are superficially similar in color and size, but they are not closely related, so you won't find them next to each other in the field guide. All birders need to become familiar with the taxonomic sequence of bird families in the field guides they use. *(left: Ohio, July; right: Manitoba, June)*

and other excellent bird guides were published, most notably, Richard Pough's *Audubon Bird Guide* in 1946, the *Golden Guide* in 1967, the *National Geographic Guide* in 1983, and the *Sibley Guide* in 2000.

Choosing a Field Guide The two authors of this book are closely associated with the *National Geographic Field Guide to the Birds of North America,* now in its fifth edition. We are proud of that guide— it covers more species than any other field guide to North American birds and is regularly updated—but we recognize the usefulness and unique qualities of other field guides currently in print. Most birders have a primary field guide and, if asked, will express strong opinions about why it's better than the others. If you are choosing your first field guide, look all of them over and read on for more advice.

How Field Guides Are Organized The best field guides do not organize species from front to back by color, or size, or habitat. Almost all field guides are organized in a similar fashion, based on taxonomy, following a scientific system of classification based on evolutionary relatedness. Similar species are grouped together because they share more recent common ancestors on an evolutionary time scale, not because they look similar. This is not an intuitive system, especially the sequencing of families from front to back *(see chapter 9)*. Become familiar with the sequence of families and know generally where they occur in your field guide to use it efficiently.

What Field Guides Do and Don't Tell You Field guides are by definition portable and compact, so valuable information sometimes has to be edited or even omitted. Most field guide authors (and experienced birders) agree on what information is most important, so there are many similarities among field guides. The main differences are in the quality and effectiveness of the visual presentation, the writing style and interests of the author, and the geographical scope of the guide.

Features The basic features found in all modern field guides are discussed below. *(See also pages 26–27).*

■ **Illustrations** Illustrations are the heart of any field guide. Although beginners are drawn to photographic guides, most experienced birders agree that painted illustrations work best. A photograph depicts an individual bird during a single instant of its life and, as a result, may show a bird in an unnatural position or with confusing shadows, backgrounds, and color shifts—making comparisons to other species difficult. A successful illustration simplifies confusing details and presents an image that better matches a birder's experience of the species in the field. As an exercise, compare photographs with artwork of the same species in several guides to understand what each type of illustration can contribute and how they complement one another.

■ **Species Accounts** Most modern field guides present text, range maps, and illustrations together on the same page or two-page spread. This is the most convenient arrangement for quick reference, but the downside is that there is very limited space for text. As a consequence, most species accounts are terse presentations of the most important field marks, including voice, with short descriptions of status and distribution. There is not room for much, if any, information on behavior, habitat, nesting, migration, or other life history details, although you will find that there are innumerable good references on such subjects.

■ **Range Maps** These thumbnail-size maps are rich in information. At a glance, you can tell where a species is regularly seen. Range maps are generalized; some species are uncommon or restricted to a specific habitat.

Chimney Swift *(left)* and **Northern Rough-winged Swallow** *(right)* are both aerial species, but they are distantly related: Swifts are in the same order as hummingbirds and so are found near them in your field guide. With a close look, notice the swift's thinner, swept-back wings, small tail, and different body shape; the leading edge of the swallow's broader wings shows an angular bend (or "wrist") not visible on the swift. *(left: Illinois, May; right: California, August)*

CHAPTER 3

STATUS AND DISTRIBUTION

One of the most important aspects of birding, relevant to birders of all experience levels, is understanding status and distribution—where and when birds are found and how common or rare they are at different seasons. Many birders are tempted to think that any species can turn up at any time and in any place. Although strays and unseasonable records do frequently appear, let us focus first on what is normal. Birds are remarkably predictable within the known and established limits of status and distribution.

When contemplating your interest in birding, it's tempting to think, "All I want to do is enjoy myself and identify a few birds." This brings you into the community of the many millions of people in North America who enjoy birding at that level. When you are ready to take the next step, however, you'll need to learn the basics of status and distribution.

Accomplished birders are in tune with the regular unfolding of the seasons and the birds that accompany them—the patterns of distribution—and have become familiar with a species' population size and habitat requirements. These birders know what to expect and what to be surprised by, or they know where to get more information. Their learning is certainly inspired by field experience, but they have a foundation of knowledge that underpins their experience.

Northern Harrier *(above)* is widespread across North America and found in a variety of open habitats. In the United States, however, **Snail Kite** *(opposite)* is now restricted to south Florida, where it inhabits large freshwater marshes. Adult males are shown. *(top: California, November; opposite: Florida, February)*

Defining Terms

Status Simply put, *status* refers to the numerical abundance of a particular species. Some of the more frequently used terms (in descending order of abundance) are *abundant, common, fairly common, uncommon, rare, casual,* and *accidental.* Most of the terminology is self-explanatory, but below we give some examples and fine-tune these terms a bit.

We all know the *abundant* species—American Robin, Common

Grackle, and European Starling across eastern North America as well as House Finch across western North America. *Common* species include Blue Jay and Northern Cardinal in the East and Mountain Chickadee in the mountains of the West. *Fairly common* species are fairly easy to find but occur in lower numbers, like Brown Creeper across much of North America. *Uncommon* species are regularly found but occur in very low numbers, like Gray-cheeked Thrush as a migrant in eastern North America. It is important to keep in mind that when a source identifies a species as common or fairly common in a certain area, this is true only in the right habitat and at the right time of year. This same species is likely to be much more uncommon, rare, or even nonexistent in a different habitat or time of year. Some species are *local,* being numerous in one area but absent in another, even when the habitat seems to be appropriate. Still other species are *irregular* (or sporadic): common to even abundant in some years yet strangely absent in others. Red-breasted Nuthatches and Bohemian Waxwings are two examples of species that wander much farther south in some years.

The birding community has made an effort to standardize the terminology for rarities. The American Birding Association and the Committee on Classification and Nomenclature of the American Ornithologists' Union have jointly embraced the following terminology: *Rare* species are those that occur in very low numbers, but occur annually, in North America. They include visitors and rare breeding

Bohemian Waxwings are unpredictable in their winter movements over parts of their mapped range, particularly the more southern locations. In most years none are found in the southernmost areas, but some years small to even moderate numbers occur, often associating with Cedar Waxwing flocks. *(New Hampshire, January)*

residents. *Casual* species are those not recorded annually in North America but for which there are six or more total records—including three or more in the past 30 years—reflecting some pattern of occurrence. *Accidental* species are those that are have been recorded in North America five or fewer times, or fewer than three times in the past 30 years. Of course, such terminology can also be applied to smaller areas.

Distribution *Distribution* refers to the range of a species—where it is regularly found. Most field guides, regional distribution works, and guides to specific bird families (such as shorebirds or thrushes) include range maps. Using different colors or patterns, a range map depicts the general breeding and winter ranges of each species *(see page 34)*. Some show the areas of regular migration, too. At a glance you can get a pretty good idea whether your own sighting is within the normal range of a species.

Range maps vary greatly in their accuracy and in the level of detail they show, and they may evolve over time. For example, it's hard to find an accurate map for Eurasian Collared-Dove, a species rapidly expanding northward and especially westward. Often, changes in distribution represent not a sudden change in a species' status but rather our increased understanding of its distribution. For instance, it is now known that Virginia's Warbler breeds in small numbers in the Black Hills of South Dakota. Yet prior to 1997, the species wasn't even recorded for the state.

Some species—such as rails and owls—are simply difficult to find; their nocturnal habits and secretive nature can mask their true status. Over the last decade, researchers have grown more adept at finding Northern Saw-whet Owls, which has altered our understanding not only of their winter range but also of their status and range during migration. As the popularity of birding increases and more birders contribute to census-type projects, our knowledge of distribution improves.

Without question, the most significant contributions toward fine-tuning our knowledge of bird distribution are breeding bird atlases, which have been completed for most states and provinces and in some instances for counties. In each atlas, a region is divided into census blocks. For some areas, atlases have been updated after 20 or so years and show us how the distribution of a particular species has changed. Some species have become more numerous and widespread; many others, unfortunately, have greatly declined. This knowledge enables us to raise the alert and develop conservation strategies for truly imperiled species.

Sophisticated census techniques and nighttime banding operations have yielded much new information about the overall distribution of the **Northern Saw-whet Owl**—both on its breeding and winter grounds as well as during migration. *(Idaho, June)*

Understanding Range Maps

One of the most important tools for understanding status and distribution is the range map. Below is a map of a hypothetical species with a key that explains all the various colors, arrows, and symbols that are used. On the next page are the range maps of two species as they appear in the *National Geographic Field Guide to the Birds of North America,* fifth edition. Use the map key to interpret their ranges. One is widespread across North America, whereas the other barely enters the United States from northwestern Mexico. In addition, the species accounts in the field guide usually give the status and habitat preferences of a species within its mapped range.

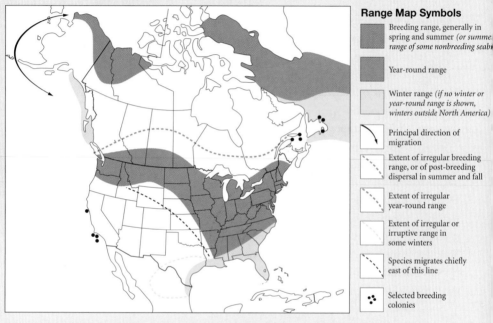

Range Map Symbols

Breeding range, generally in spring and summer (*or summe range of some nonbreeding seab*

Year-round range

Winter range (*if no winter or year-round range is shown, winters outside North America*)

Principal direction of migration

Extent of irregular breeding range, or of post-breeding dispersal in summer and fall

Extent of irregular year-round range

Extent of irregular or irruptive range in some winters

Species migrates chiefly east of this line

Selected breeding colonies

Map of a hypothetical species.

The Importance of Status and Distribution

The excitement of local birding is greatly enhanced by watching for new arrivals and departures through the seasons and by keeping track of the permanent and seasonal residents. Through your own discoveries, interactions with others, and reading, you will become familiar with the status and distribution of your local birds.

Begin at Home Start with an area that is well known to you—most likely your own yard—and then work outward. Keep notes of what you see and record numbers. By making daily recordings, you will begin to learn which species are always at hand, which are present only in summer or winter, and which are present only during migration time.

Belted Kingfisher is a partially migratory species that is widespread across North America. Because it plunge-dives for prey, it withdraws from the northern portion of its breeding range, limited by its need for open water. Kingfishers are quite local during summer: They need streams or rivers with sandy banks for the excavation of their nest tunnels. (Connecticut, February)

Five-striped Sparrow—one of the highly patterned Mexican sparrows in the genus *Aimophila* that barely enters the U.S.—is found in a few steep, narrow canyons of southeast Arizona with brush and cactus-covered hillsides and year-round water. In North America, it is best known from California Gulch, west of Nogales. Note the five white stripes. (Arizona, August)

Keeping track of species' arrival and departure dates is also a great way to learn. The dates won't be the same from year to year. The surprise of recording a new species for your yard or a surprisingly early or late date of a species' arrival is part of the fun of birding.

Expand Your Horizons Most birders also have their own favorite local patch, a nearby place with lots of birds, where they regularly go birding. It might be a patch of woods with a stream that is particularly exciting during migration, or it might be a pond or wetland that attracts waterbirds. When birding your patch, use the same techniques you use for your yard: Take notes and keep track of which species you see, their numbers, and the dates. When you're ready, look at the broader picture—perhaps starting with your county and then eventually your state or province.

Draw Conclusions As you take notes, you will find that some birds are *resident*—present year-round—like Northern Cardinal over so much of the East or House Finch in the West (and much of the East, where it was introduced). In reality, few species are totally sedentary; many resident species move around locally and, more rarely, disperse over large distances.

Two western species that barely move at all are Wrentit and California Towhee. The sedentary Wrentit is unrecorded from Washington, even though it is fairly common just over the Oregon border on the south side of the Columbia River. Other species are *winter visitors,* spending only the colder winter months in your yard. Widespread winter visitors in the East include White-throated Sparrow and Dark-eyed ("Slate-colored") Junco. White-crowned Sparrows (of the *gambelii* subspecies) are common winter visitors in much of the West. Still other species are *summer visitors,* like Baltimore Oriole across much of the East or Bullock's or Hooded Orioles in parts of the West. Some species that you see in your yard will be strictly *migrants* or transients that appear only in spring and fall. Swainson's Thrush is a widespread North American species that appears in most yards only as a migrant.

Adding to Your Understanding of Status and Distribution

Once you've gathered your data and made some initial conclusions, you'll want to check the validity of what you've found and learn more about habitat and rarities.

Review with Your Mentor Another way to learn status and distribution is to consult more experienced birders. Having a close mentor is particularly helpful. You'll want to ask whether your particular sighting makes sense and whether it fits the known patterns. If you live in the Midwest and you have just seen an adult male Baltimore Oriole in your yard in late April, you are likely to hear, "Congratulations," and "Wasn't it beautiful?" If you glimpsed what you think may have been a Baltimore Oriole on a cold March day, you may be asked, "How well did you see it?" or "Could it have been an American Robin?"

As with other aspects of birding, always ask questions and learn from your mistakes. All birders, even the experts, know that we learn *most* from mistakes. Those who bird alone are at a disadvantage in this regard, as they don't have anyone to check their identifications.

Learning Arrival and Departure Dates

This page of bar graphs is taken from *Birds of Southern California: Status and Distribution,* which was published in 1981.

Key to Districts: C: Coastal; M: Mountains; D: Desert; S: Salton Sea; R: Colorado River

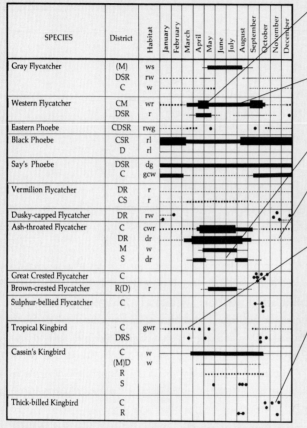

Common to Abundant: Almost always encountered in proper habitat within the given district(s), usually in moderate to large numbers.

Fairly Common: Usually encountered in proper habitat at the given season(s) in the given district(s), generally not in large numbers.

Uncommon: Occurs in small numbers or only locally under the indicated conditions.

Rare: Occurs annually, but in very small numbers. Also includes species that breed extremely locally in very small numbers.

Casual: Records within a given district in the season indicated are few, but not cited individually. May be no records in some years, but a general pattern of occurrence is suggested.

Individual Record: Details in text. Generally includes species with fewer than ten records in the region, or recorded only a few times within a district or in that season. For individual that remained for an extensive period, dots are connected by a dashed line.

Consult Publications In addition to asking your mentor and birding friends lots of questions, you should have a birding library that includes not only references on identification but also works on the status and distribution of the birds of your area. You will find publications addressing species distribution for most states and provinces; you may even find works and checklists specifically for your county or region. Many of these works have seasonal bar graphs that facilitate learning arrival and departure dates and when peak movements occur *(above).*

Some poor-quality publications dispense little information of value. The worst are those works that include many erroneous records—just the sort of information you shouldn't learn! To avoid publications like

Le Conte's Thrasher is found locally across the somewhat open deserts of the Southwest from the San Joaquin Valley in California to southern Nevada and southern Arizona. It is often seen along dry washes and is frequently encountered running. Listen for its loud upslurred whistle. *(California, March)*

these, check popular birding magazines for reviews of new publications and ask your mentor or more experienced birders in your area (or the area in which you'll be birding) which works are reliable.

In general, we feel that the authors of distributional works should be extra critical when choosing which records to include. As two birders who are preparing their own exhaustive distributional work on the birds of Inyo County, California, put it, "Do no harm." They were referring to authors who obfuscate the truth by including erroneous information.

There are hundreds of outstanding works on status and distribution. Particularly worthy of note is the recent *Birds of Hamilton and Surrounding Areas*. This work is confined to a small area in Ontario, but the information is so carefully compiled, using more than a century of ornithological data, that it is a vital reference for anyone birding in southern Ontario or the adjacent United States.

Regrettably, some classic works are out of print, but you may be able to find used copies of these publications. Referring to a good distribution work on your area or on adjacent regions will greatly facilitate your understanding of status and distribution. *(See also "Additional Reading" on page 218.)*

Become Familiar with Habitats In addition to learning the ranges and distributional timing for an individual species, it is also important to learn its habitat—the ecological niche where it is found. Knowing the precise habitat is essential for finding local species during the breeding season. Even during the winter and migration periods, most species frequent particular types of habitats. Along the Atlantic coast during winter, one's search for Purple Sandpiper will be confined to rocky coastlines (including jetties) rather than sandy beaches. A clear tumbling forest stream in the East is likely to be the place where Louisiana Waterthrush breeds, while Northern Waterthrush prefers the slower moving waters in bogs. Le Conte's Thrashers favor open desert with sparse, low vegetation and lots of room to run—they often frequent washes within such habitat—whereas the closely related Crissal Thrasher prefers more heavily vegetated sites, especially those near watercourses and in stands of mesquite or pinyon-juniper. Knowing what type of field might contain breeding Henslow's Sparrows will greatly enhance your success in finding this highly sought-after, scarce, and local species.

Know What to Do When You Find a Rare Bird When a Ross's Gull—usually found in the far north, outside North America—appeared in early 1975 in Newburyport, Massachusetts, the dean of birders, Roger Tory Peterson, called it the "bird of the century." Yet now that we have a few dozen additional records, there is an established pattern for Ross's Gull in the lower 48. Such a rare sighting is exciting, but few records stand alone for long. Sooner or later another one appears. American Three-toed Woodpecker, native to the boreal and western forests, made a surprising appearance recently in western Kansas—on the Cimarron National Grassland (Morton County)—in midsummer (early July 2005). Yet as remarkable as this record was, it turns out that there were

This male **Bobolink** is a striking sight in a flowering field. Breeding Bobolinks require large open grassy areas, often using clover-alfalfa hayfields. They breed locally across much of the northern United States and southern Canada. (New York, June)

two earlier well-documented records from Nebraska (June 1916 and July 1994), also in summer.

Over time, the occurrence of rarities forms a pattern. A rarity could be a species rare at any time for your area, an early or late record of a more common species, or a nearby resident that is slightly out of range. To be well informed, become familiar with the distribution patterns and seasonal timing for both more common and unusual species. As 19th-century scientist and physician Louis Pasteur once said, "Chance favors the prepared mind." Being prepared—knowing which species are regular in your area and when regular migrants arrive and depart— will enhance your overall competence as a birder. Expert birders who must review exceptional records, such as members of committees that review state or provincial records, sometimes hear, "I had no idea my sighting was unusual." If the birder didn't know the sighting was unusual, then he or she didn't look at the bird critically, and it is thus much more likely a mistake was made. A good birder will remember these mishaps and learn from them. If you are still convinced of the accuracy of your sighting, submit documentation (a written description including photographs or drawings, if you have them) to those in charge of reviewing records from your region.

The Fascination of Learning Distribution

As you delve into the study of distribution, the pieces of the puzzle will start to fall into place. You'll find that while some species are almost

Pectoral Sandpiper is one of the most migratory species in the world in terms of the distances traveled. The adults *(below)* migrate through the Ohio and Mississippi river regions in both spring and fall. Later in fall, the more colorful and freshly patterned juveniles *(opposite)* migrate south on a broader front; small numbers of juveniles can be found on both the West and East coasts. *(below: Ohio, August; opposite: Ohio, September)*

entirely sedentary, with little change in distribution and abundance through the seasons, others show complex patterns of distribution and abundance through the seasons.

Consider, for example, Pectoral Sandpiper, a shorebird that breeds in the high Arctic and winters in southern South America. In spring, adult Pectorals mostly migrate across North America in a broad path through the Ohio and Mississippi river valleys. They arrive on the Texas coast in early March and reach their Arctic breeding grounds in late May or early June. Their breeding range extends from central Canada west to Alaska and the Russian Far East (even as far west as the Taymyr Peninsula on the central Russian Arctic coast). In fall the adults essentially retrace their spring route, arriving in the Ohio and Mississippi river valleys by early July. Their fall migration continues on into November. (In nearly all shorebirds, adults move southward first, followed by juveniles.) For birders in both the East and the West, Pectoral Sandpiper is known as an uncommon fall migrant, arriving in late August with peak numbers from mid-September to mid-October. We find differences in timing from these geographical regions because outside the main Midwest migration path, the overwhelming majority are juveniles, and they aren't seen south of the Canadian border until late August.

Many birders would agree that warblers as a group are the most compelling birds that migrate through North America, and Connecticut Warbler may be the most highly sought-after species among them. Rather scarce overall and secretive, Connecticut Warbler has the most

poorly documented winter range of any of our wood-warblers. It probably winters south of the Amazon Basin in South America. Its arrival in early May (earliest record is April 27) in south Florida establishes it as our latest-arriving wood-warbler. It typically migrates west of the Appalachians in spring, arriving in the upper Ohio and Mississippi river valleys after May 10 and in the southern Great Lakes after May 15. Yet there are many purported sightings of Connecticut Warbler in late April and the beginning of May in the upper Midwest and upstate New York.

Are these latter sightings really incorrect identifications? They probably are, for two reasons: First, all are undocumented, and, second, considering that the earliest record for south Florida is April 27, earlier records for upstate New York are unlikely. In fall, much of Connecticut Warbler's migration is over water, well off the Atlantic coast, on its return to South America. Proof of a substantial transatlantic flight came on September 26, 1987, when a remarkable 75 Connecticut Warblers were recorded in Bermuda following the passage of Hurricane Emily—twice the total seen in the last several decades!

Of course, unseasonable records do occur. In spring in the East, unusually early sightings often follow the arrival of strong warm fronts with powerful southwest winds. The passage of such fronts (one in 1947 was particularly notable) has brought such early migrants as Blue-winged and Hooded Warblers as far north as the Great Lakes by early April. But note that migrants of these species had *already* arrived on the Gulf Coast. Since the passage of these fronts is usually followed by a strong cold front with subfreezing temperatures, the survival of most of these

These two look-alike species of vireo are not closely related. **Plumbeous Vireo** (left) is widespread in the Rockies and the Great Basin ranges. **Gray Vireo** (right) breeds in brushier regions at somewhat lower elevations. While Plumbeous is regularly noted in migration, Gray Vireo is virtually unrecorded away from known nesting and wintering areas. Note the much longer tail of Gray Vireo and its lack of white spectacles; it also has very short primary projection past the longest tertial. (left: New Mexico, May; right: New Mexico, June)

early arrivals is in doubt. Establishing and documenting these record-setting early and late dates is part of the challenge and fun of birding.

Another example from western North America involves the status of Cassin's and Plumbeous Vireos. Both are part of the Solitary Vireo species complex and were considered (along with the Blue-headed Vireo from eastern North America) a single species until recently. Cassin's Vireo has an extended spring migration through the lowlands, with a few birds arriving as early as late March. Its fall migration is also early, primarily mid-August to late September. Plumbeous Vireo is a later spring migrant, not arriving in eastern California until mid-May (earlier in the Southwest) and remaining on its breeding grounds until late September. In the fall, most migrant Plumbeous Vireos are noted in October, after Cassin's Vireos have departed. In the East the timetable of Blue-headed Vireo (which is very similar in appearance to Cassin's) is different from that of both Cassin's and Plumbeous. Blue-headed is an early spring migrant (like Cassin's) but a late fall migrant (like Plumbeous). And what of Gray Vireo, which, though similar in appearance to Plumbeous, is genetically very different? In 40 years of birding, we have never seen a migrant Gray Vireo. As noted ornithologist Philip Unitt has postulated, most Gray Vireos apparently fly nonstop from their winter grounds to their breeding grounds and then from their breeding grounds to their winter grounds. There are very few well-documented records of Gray Vireos during migration, yet there is a remarkable early October specimen record from Wisconsin.

For some *polytypic* species—species with multiple subspecies *(see page 210)*—the distributional patterns can be remarkably complex. Take, for instance, part of the Pacific subspecies of Orange-crowned Warbler, *lutescens*. Those that breed on the west side of the Sierra Nevada from late March through early May migrate upslope in late May to montane meadows to spend summer. Some overshoot and appear at desert oases at a time when birders are looking for late spring migrants and eastern vagrants. Yet in the boreal subspecies (nominate *celata,* the only one occurring regularly in the East), the fall migration is principally in October. Indeed, the earliest reliable record from well-birded Point Pelee in Ontario is September 17. So in a broad sense, then, we can say that the start of fall migration for the Orange-crowned Warbler is from late May to mid-September, depending on the population or subspecies involved.

This chapter has illustrated both how interesting distribution is and how your birding is enhanced by a keen grasp of distribution and status. Having considered the big picture, we look in the next chapter at the smallest details—the parts of the bird.

CHAPTER 4

PARTS
OF A BIRD

K nowing the parts of a bird—the names and locations of feather groups and bare parts—is important to the identification process. They are an element of the "language" of birding, and most field guides assume you already have a basic understanding of them. All the names and details may seem overwhelming when you're just starting out. Don't worry, though—you can still enjoy going birding and you can identify many birds even if you don't have all of the terminology memorized. You don't have to learn everything before you start; over time and with experience, you will become fluent in this new language.

So dip into this chapter and refer back to it later as needed. Read the text and captions, and then study the photographs and drawings. Many of the illustrations and images in this chapter have been annotated extensively—labels have been placed directly on the images—in order to distinguish specific parts of a bird or to indicate differences between similar birds.

Another good learning technique is to look at unlabeled photographs and name the field marks you see. In the next chapter, "How to Identify Birds," you'll see how classic field marks are used in addition to details of size, shape, and behavior to solve specific identification challenges.

The "ears" on this **Long-eared Owl** are ornamental feather tufts; the ear openings are located slightly below the eyes and behind the rufous feathers of the facial disk. The composite image of Long-eared Owl *(opposite)* features a bird from a winter roost site used by over 40 birds. In flight, the ear tufts are flattened against the head. *(both: California, January)*

Bird Topography and Identification

Knowing the parts of a bird—*bird topography*—will help you to understand how a bird's body is organized and why it is shaped the way it is. This knowledge will guide you to the right field marks when you are birding in the field.

Feathers and Feather Groups The evolutionary development of the feather is shrouded in time and mystery. What we do know is that

All the stripes on the head of this **Clay-colored Sparrow** are labeled. Compare this species to the subtly different winter Chipping Sparrow opposite. (Manitoba, June)

lateral crown stripe
supercilium
postocular stripe
auriculars
nape
median crown stripe
supraloral
lores
malar stripe
submoustachial stripe
moustachial stripe

feathers brilliantly solve a number of survival issues: flight, insulation, camouflage, communication. Overlapping feathers cover almost entirely the exposed areas of a bird's body. The unfeathered parts—bill, facial skin, eyes, legs, and feet—are referred to as the *bare parts*. Some bare parts have special terms, which will be covered in this chapter.

Different types of feathers have different functions, but most feathers are built around a basic plan. The central feather shaft has two flat *webs* (or *vanes*) of barbs along its upper length that are knit together by barbules featuring Velcro-like hooks and edges. If a feather becomes disheveled, the interlocking quality of the barbules allows a feather to be preened and the barbules reknit, something most of us have experienced by stroking a feather between our fingers from its base to its tip.

A bird's feathers don't cover its body uniformly, unlike fur on mammals. They grow in symmetrical groups that have a precise arrangement, often visible as discrete, overlapping rows. Most feather groups are found in all families of birds, and the terms used to describe them are the same for all species. Birds as different as condors and chickadees each have scapulars, primaries, and uppertail coverts. They may have different shapes, sizes, and colors, but typically the same feather groups are found in the same relative positions. This chapter starts with examples from the passerines and then moves on to other groups of birds.

Passerines

The *passerines* are a large order of birds that includes all our songbirds—thrushes, warblers, sparrows, and such—as well as the slightly different tyrant flycatchers. There are more than 400 species of passerines in over 30 families recorded from North America; they fill the back half of your field guide. Their body plan and feather organiza-

eye line: extends to the bill and includes the dark lores

lores

no moustachial stripe

The labels point out the features of this winter **Chipping Sparrow** that differ from those of Clay-colored Sparrow *(opposite)*. The overall effect of the dark lores and eye stripe gives Chipping Sparrow a "stern-faced" look different from Clay-colored. *(Texas, January)*

tion are rather simple and uncomplicated. For this reason—and because passerines include many of our most well-known species—this is a good place to start a discussion of bird topography.

The Head There are a number of named feather groups on the head. The patterns they form are important to identification; a careful look at the head can often be enough to provide a correct identification. Clay-colored Sparrow *(opposite)* has contrasting patterns on most of the important feather groups of the head. Sparrows are a confusing group, and their head markings help to sort them out. Some species are not as conspicuously marked on the head—for example, Pine Grosbeak *(see page 50)*—but the feathers still form groups that can be distinguished. Look for seams between the feather groups; these will show up as shadows or faint lines. The discussion that follows starts at the top of the head and works down.

■ **Crown** The *crown* is the area on top of the head, from the bill to the back of the skull. The short crown feathers often subdivide into noticeable contrasting rows that have their own names. At the centerline of the head, Clay-colored Sparrow has a white stripe, known as the *median crown stripe,* bordered on either side by dark *lateral crown stripes.* On some species, like Northern Cardinal or Pyrrhuloxia, the *rear crown feathers* are elongated into a *crest.* Other species—such as kingbirds and kinglets—have a *crown patch* of a contrasting color that is typically hidden by the surrounding crown feathers.

■ **Supercilium** The feather group comprising multiple rows of tiny feathers below the crown that extend from the bill to behind the eye is called the *supercilium.* It is also known as the *eyebrow,* especially when it is a contrasting color. The supercilium of Clay-colored Sparrow is uniformly pale. The width and color of the supercilium can vary. The part of the supercilium between the eye and the upper mandible is known as the *supraloral.*

■ **Lores and Facial Stripes** The area between the eye and the bill, known as the *lores,* is usually covered with tiny feathers. Whether this area is dark or pale can be significant: On Clay-colored Sparrow it is pale, and this immediately separates it from all Chipping Sparrows, which have dark lores.

The discussion of the various facial stripes that follows may seem complicated, so keep referring to the photograph of Clay-colored Sparrow *(see page 46),* where all these terms are labeled. If it feels like too much to absorb, read it through once and come back to it when you can apply it to a bird identification problem you've experienced.

Below and behind the eye is a specialized group of feathers known as the *auriculars*; they are so named because they cover the ear (aural) opening. This area is also referred to as the *ear coverts* or *cheek.* The central feathers of the auriculars are open, lacy feathers that allow sound to pass easily. The feathers around the perimeter of the auriculars are heavier and often form contrasting lines and spots that have different names. Notice how Clay-colored Sparrow has a dark line behind the eye and just above the auriculars; this is known as a *postocular stripe.* By contrast, some species with dark lores, such as Chipping Sparrow,

This adult **White-eyed Vireo** has yellow spectacles that contrast with the rest of its gray head. The parts of the bill are also labeled. *(Ohio, May)*

spectacles: yellow wraps around eye and connects to supraloral

white iris

black pupil

culmen

upper mandible (maxilla)

lower mandible

gape

have a line that extends through the eye to the bill, known as an *eye line* or *transocular line*. On Clay-colored Sparrow, the lower border of the auriculars forms a contrasting line that starts at the gape of the mouth; this feature, known as a *moustachial stripe,* is absent on all Chipping Sparrows. Below the moustachial stripe is a broader, pale stripe called the *sub-moustachial stripe,* and below that is a fainter, dark stripe called the *malar stripe.* Other terminology has been proposed for these facial stripes, but the terms given here have long-standing acceptance in both North America and Europe. Behind the crown and auriculars is the *nape* (or back of the neck). On Clay-colored Sparrow, this is prominently gray.

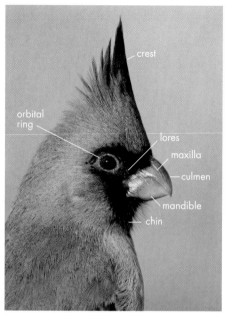

A southwestern relative of Northern Cardinal, **Pyrrhuloxia** has a striking crest and a bill with a strongly curved culmen. This is a male; females have very little red in their plumage. *(Texas, December)*

■ **Eyes and Eye Feathering** Bird pupils are always black, but the *iris* can be colored. References to eye color in your field guide refer to the color of the iris. Eye color is an important field mark in some species: For instance, female Bushtits have pale eyes, whereas males of this species have black eyes. Eye color can also change with age (juvenile White-eyed Vireos have dark eyes) or with the time of year (adult Brown Pelicans have eyes that change color on an annual cycle; the pale eyes seen in late winter and spring darken by late summer).

The skin that surrounds the eye is known as the *orbital ring.* The orbital ring is rarely prominent on passerines, but its color is useful for identifying some gulls and a few other species, such as cuckoos. An important feature on some passerines is a noticeable *eye ring* formed when the tiny feathers that grow in rings around the eye contrast with the surrounding feathers. Various patterns of eye rings can give a species its distinctive look. A complete eye ring forms a circle around the eye. When a complete eye ring widens behind the eye, it forms a teardrop shape. When an eye ring is interrupted in front and behind the eye it is called a broken eye ring. An *eye crescent* (or eye arc) is a partial eye ring that may be visible above the eye, below the eye, or in both places. *Spectacles* are formed when a pale eye ring connects with a pale supraloral area and contrasts with the rest of the face. The orbital ring and eye ring often get confused with each other: Remember that orbital rings are made of skin, eye rings of feathers.

■ **Bill and Bill Feathering** The size, shape, and color of the bill can be important field marks. The upper half of the bill is known as the *upper*

Every wing feather on this **Pine Grosbeak** is beautifully edged in white, and the various feather groups are clearly delineated. The primary projection is the distance from the tip of the longest tertial to the tip of the longest primary. (Michigan, January)

mandible or maxilla. The upper edge of the maxilla is called the *culmen,* and the shape of the culmen (from arched or curved to straight, or even recurved) can be an important field mark. The lower half of the bill is known as the *lower mandible*; alternatively, it is referred to simply as the *mandible* when maxilla is used for the upper mandible. The *gape* is the point where the upper and lower mandibles meet. On most passerines, the feathers at the base of the upper mandible hide the nostrils. Specialized, bristle-like feathers that extend forward from above the gape, known as *rictal bristles,* are found on many species, but these are rarely important for identification. Other species have elongated *nasal bristles* that cover the nostrils and extend forward onto the bill. This feature can occasionally be an important field mark: On Chihuahuan Raven these feathers cover more of the bill than those on the very similar Common Raven.

The Outer Wing The wing is a geometrically organized system of feathers honed to perfection by the evolution of flight. The basic wing structure is similar in most birds, but the shapes and lengths of the feather groups vary considerably. In passerines the wing is most often observed in the folded position because these smaller species rarely soar (although ravens do) and their wings flap too quickly for detailed study. Even so, it is very useful to know both how the wing opens and closes and how the various feather groups overlap and change position. The next time you come across a freshly dead bird—one that has hit a window, perhaps—take the opportunity to examine it in detail. By opening and closing the wing, you can study how the wing is put together. (In a situation like this, use common sense: Wash your hands after you examine the bird, and remember that with few exceptions it is illegal to possess any wild bird, even a dead one, or its feathers.) While reading this section on wings, refer to the labeled photograph of Pine Grosbeak *(opposite)* and the illustration of Lark Sparrow's wing *(below)*.

The next few bulleted entries refer to features on the outer wing or "hand." The all-inclusive term *flight feathers* describes the major feathers of the wing—primaries and secondaries—in addition to the tail feathers. Some birders use the term *remiges* (singular, *remex*) to describe the flight feathers of the wing. The *bend of the wing,* the *wrist,* and the *carpal* are all names for the joint between the outer wing and the inner wing. On perched birds, overlaying body feathers alongside the breast often conceal the bend of the wing.

■ **Primaries** The *primary feathers,* so called because they attach to the first or outermost bones of the wing, are the large flight feathers of the outer wing. Most species have nine or ten functional primaries and a tiny outer primary *(remicle)*. In some species—including corvids, wrens, and thrashers—the outermost functional primary is much shorter *(see photograph of Sage Thrasher, page 53)*. This makes for a more rounded wing tip, which is helpful to birds that fly in brushy areas. Long-distance migratory species tend to have longer primaries and tails than those found on related shorter-distance migrants; longer flight feathers are more efficient on longer flights. Each primary feather has a *primary number* that helps to pinpoint its location. In North America we number the primaries from the innermost (p1) to the outermost (usually p10), as illustrated.

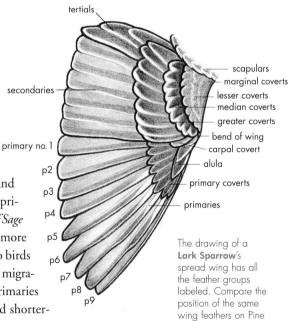

The drawing of a **Lark Sparrow**'s spread wing has all the feather groups labeled. Compare the position of the same wing feathers on Pine Grosbeak *(opposite)*.

On passerines, identification features involving the primary feathers typically have to do with feather proportions and length rather than color or pattern. For the most part, the primaries are hidden when the wing is folded because they slide under the secondaries and tertials. The tips of the folded primaries, which form the *wing tip,* are often visible beyond the tertials. The *primary projection* (or primary extension) is the distance that the wing tip extends past the longest tertial feather; as you can see, it is rather long in Pine Grosbeak. In some groups—many pipits, for example—the primaries are completely hidden by the overlaying tertials, so there is no primary projection. In a very few cases, *primary tip spacing,* the gaps or spaces between the tips of the folded primaries, can be a critical aid to identification.

■ **Primary Coverts** As a general term, *coverts* refers to the small feathers arranged in rows that overlay any of the flight feathers (primaries, secondaries, and tail). The *primary coverts* overlay the bases of the primaries. If there are multiple rows of coverts, the outermost row is referred to as the *greater primary coverts,* the next row as the *median primary coverts,* and the next as the *lesser primary coverts.* In most passerines, the primary coverts have only one row of visible feathers and they are rarely prominent; on Yellow-headed Blackbird, however, the primary coverts form a white patch on its otherwise black wings.

■ **Alula** The feathers of the *alula* (sometimes referred to as the "bastard wing") attach to a small bone near the bend of the wing and overlap the primary coverts. The stiff alula feathers, usually three or four

Swallows are the most aerial group of songbirds. This adult **Barn Swallow** has a long forked tail: The outer tail feathers are much longer than the central tail feathers. The aerodynamic "swallow-tailed" shape allows the bird to make very tight turns and adds lift; the long pointed wings allow it to glide slowly and reduce flapping—a perfect combination of grace and efficiency. (California, May)

wing linings include
the primary and
secondary coverts

p10

A **Sage Thrasher** in threat display allows a good look at the under-wings. The feathers of the secondary under-wing coverts are loose and disorganized. On passerines, all you can hope to see is the general color of the wing linings. The short tenth primary (p10) is full-grown. (New Mexico, June)

in number, function during a variety of flight actions. They help prevent stalling at low speeds by being extended to form an extra feather slot. This is sometimes visible on hovering raptors, for instance. They are of minimal importance on passerines.

The Inner Wing The *inner wing* includes all the feather groups between the bend of the wing and the body. In the passerines, these groups include the *secondaries, tertials,* and *secondary coverts.*

■ **Secondaries** The secondaries are the long flight feathers of the inner wing or arm. They slide over and cover the primaries as the wing folds. Most passerines have six secondaries (not counting the three tertials); long-winged species such as pelicans can have up to 30 or more secondaries. The tips of the secondaries and primaries form the *trailing edge* of the spread wing.

■ **Tertials** These large feathers, known as *tertials,* are the innermost secondaries, but because they differ in shape and function, they are considered as a separate group. On passerines there are usually three tertial feathers; larger birds may have more. They are graduated in length; that is, the inner (first) tertial feather is much shorter than the outermost one. The tertial feathers are prominent feathers on many species, often with distinctive colors and patterns that can be important for identification.

When the wing folds, the tertials cover and protect the secondaries stacked beneath them. On the folded wings of most passerines the secondaries are barely visible—as a stack of lines—beneath the prominent tertials and above the primaries.

■ **Greater Coverts** Covering the bases of the secondaries and tertials

These two images of **Smith's Longspur** in nonbreeding plumage show the same individual—a casual visitor to California. Try to visualize the wing folding and unfolding, and follow the location of the missing greater covert feather. When the wing is closed, some feather groups are hidden. The diagnostic tail pattern is not easily seen on longspurs, but the lower image shows it well because the tail is spread. The white pattern on the two outer tail feathers (r5 and r6) confirms the identity of this bird. (California, October)

and extending to the front of the wing are the *secondary coverts,* arranged in a series of overlapping rows of decreasing size. The largest row of coverts, the *greater coverts,* is located just above the secondaries. (For brevity, the modifier *secondary* is dropped when referring to any of these rows.) There is usually one greater covert associated with each secondary feather. The *inner greater coverts* are the greater covert feathers nearest to the scapulars or closest to the body when the wing is open.

■ **Median Coverts** The next row of shorter coverts that overlays the greater coverts is referred to as the *median coverts.* Many passerine species have pale tips on the greater and median coverts that form *wing bars.* When two wing bars are present, they are referred to as the *upper wing bar* and the *lower wing bar.* On the folded wing, the upper wing bar formed by the median coverts is often partially hidden by surrounding feathers.

■ **Lesser Coverts** Above the median coverts, the *lesser coverts* form multiple rows of small feathers; how many rows depends on the size of the bird. These feathers are less neatly organized and often do not form precise rows. Very small, stiff feathers known as *marginal coverts* protect the forward or *leading edge* of the wing. These feathers are usually hidden when the wing is folded.

The Underwing The underside of the wing is only visible when a bird is in flight or when a perched bird stretches. It is important to know that the two sides of the same wing often have different patterns and coloration. Although the same primary and secondary feathers are visible from both above and below, their different surfaces can look different.

The *underwing coverts* and upperwing coverts are separate feather groups. On passerines the underwing coverts are less extensive than the upperwing coverts, and the various rows of feathers blend together. As a group, the underwing coverts are also known as the *wing linings*. A separate group of feathers covers the area where the underwing attaches to the body; this group is called the *axillaries* (or *axillars*). Sometimes birders refer to these feathers as the wingpit. Underwing features are much more important and visible in many nonpasserines, especially in large soaring species, such as raptors and seabirds.

The Tail The arrangement of the tail feathers and their corresponding coverts is straightforward. Most species have 12 tail feathers or *rectrices* (singular, *rectrix*) arranged in symmetrical pairs. Some nonpasserine species have more tail feathers; a few groups have fewer tail feathers: Anis have eight, hummingbirds ten; and the tail feathers are vestigial on grebes, replaced by a downy tuft.

Tail feathers come in myriad shapes and sizes. Compare, for instance, Scissor-tailed Flycatcher, which has a tail longer than its body, to the stub-tailed Winter Wren. What you see of the tail's pattern and color can change with a bird's posture or activity—and with your viewing position. Important field marks are often located on the tail, especially the outer tail feathers. To observe these important marks properly, you need to understand how the tail folds.

■ **Tail from Above** Looking at the photograph of Great Crested Flycatcher with its tail open, it is easy to imagine the tail as a fan *(page 56)*. If you are looking at the tail from above, as the fan closes, the two *central tail feathers* overlay the pairs of outer tail feathers. The tail

When this **Great Crested Flycatcher** folds its tail—the normal position—all the rufous color on the outer feathers will be hidden beneath the dark brown pair of central tail feathers (r1). Notice how the rufous inner webs are much wider than the dark outer webs on the outer tail feathers (r2–r6). To see the diagnostic pattern on the folded tail you must observe it from below, as in the photograph *(opposite)*. The broad edges of the innermost tertials are a secondary field mark. *(Ontario, June)*

uppertail coverts

broad, sharply contrasting edge of the inner tertials

r6

r5

r4

r3

r2

r1

r1

r2

r3

r4

r5

r6 outermost tail feather

central tail feathers

feathers are given numbers: Both feathers of the central pair are labeled r1 (*r* is for rectrix). The other pairs are numbered in ascending order: For a species with 12 tail feathers, the outermost tail feather on either side is r6.

If you are looking at the upper side of the tail in its normal, closed position, only the central tail feathers will be visible. On most species this view is uninformative because the central tail feathers are often dark and unpatterned. You will need to work yourself into position to see the underside of the tail, for this is where the most useful details are found.

■ **Tail from Below** The second photograph of Great Crested Flycatcher shows its tail as seen from below and in its normal, closed position *(opposite)*. When observing passerine tails, try to view the tail from below, in its closed position. The two tail feathers visible in this view are the undersides of the outermost pair (r6) with one overlapping the other.

In the genus *Myiarchus,* which includes Great Crested Flycatcher, the pattern of the outer tail feathers is a critical field mark. The pattern of rufous and dark brown is clearly visible from this vantage point; note the differences evident in the photograph of Ash-throated Flycatcher *(below).* The pattern can be also seen from above when the tail is widely spread, but views of spread tails from above are usually very brief and difficult to assess. This is what you need to remember for passerines: On the closed tail from above, the upper sides of the two central tail feathers (r1) are visible; from below, the underside of the outermost tail feathers from both sides (r6) are visible.

■ **Tail Shape** The relative lengths of the tail feathers determine the shape of the tail. When looking at the underside of the tail, a close look at the tip of the tail will often reveal the relative lengths of the individual tail feathers.

On Great Crested Flycatcher, all the tail feathers are about the same length, giving the tip of the closed tail a squared-off shape with just a small fork in the center. On that species the pattern and color of its tail are more important than its shape. *Forked tails* result when the tail feathers get progressively longer toward the outside, creating a notch in the center; Scissor-tailed Flycatcher and Barn Swallow are extreme examples. The presence or absence of a tail fork separates House Finch (with its squared-off tail) from the similar Purple Finch (which has a

These two similar fly-catchers are both in the genus *Myiarchus.* Note the diagnostic difference in the pattern of brown and rufous on the outer tail feathers (r6). On **Great Crested Flycatcher** *(left)* the brown is restricted to the outer edge of the tail; on **Ash-throated Flycatcher** *(right)* the brown is more extensive and covers the entire tip of the inner web. *(left: Texas, June; right: California, July)*

rufous on inner web extends to tip

feather shaft

brown covers entire tip of inner web

From behind and with its wings drooped, the upperparts of this **White-crowned Sparrow** are fully visible. This individual (subspecies *gambelii*) has a pale supraloral and an orange-yellow bill. The streaked back feathers contrast with the brown rump, the duller scapulars, and the gray nape; most of the secondaries are hidden by the tertials. *(New Mexico, November)*

nape
back
scapulars
rump
greater coverts
tertials
longest uppertail coverts
wing tip
tail

forked tail). On graduated tails, the central tail feathers are the longest and the outer tail feathers are progressively shorter; cuckoos, shrikes, and towhees have graduated tails. When you look at the closed tail from below, you will see the tips of the individual feathers, how they are spaced, and any pattern there. This is helpful for separating similar birds such as Yellow-billed and Black-billed Cuckoos.

■ **Tail Pattern** A number of terms are employed to describe tail patterns. Pale tips on the outer tail feathers can form *tail spots*. On some species the white tail spots are located farther up the tail and the tips of the tail feathers are dark. This pattern—which can be seen on many wood-warblers—varies from species to species. A number of passerines have extensively white outer tail feathers that show in flight, especially when landing. The white pattern on the outer tail feathers of longspurs is diagnostic but requires careful study *(see page 54)*, whereas the white flash on a Dark-eyed Junco's outer tail feathers quickly confirms its identification in flight. Tail bars (or bands) that run across the tail, rather than lengthwise, are uncommon on passerines but very important in the identification of flying raptors.

The Upperparts All the feathers that cover the body are relatively small, and the feather groups blend into each other at their edges. On the upper body the four main group are scapulars, back, rump, and uppertail coverts.

■ **Scapulars** The group of feathers that originates from the small area of skin where the wing attaches to the body (and where the scapula bone or shoulder blade is located) is called the *scapulars*. The scapular feathers overlay and partially cover the inner wing, smoothing the connection between the wing and the body. Scapulars are not prominent on passerines and tend to be duller and less conspicuous than the back feathers; the division between the two groups is sometimes hard to see. In other groups of birds, the scapulars are well developed and often important for identification. They are discussed in more detail in the shorebirds section *(see page 74)*.

■ **Back** The *back* is located below the nape and between the left and right scapulars; the folded wings never cover this area. The back feathers, which are moderately long, are arranged in rows. Various patterns can occur on the back, from solid colors to intricate, multicolored stripes and fringes. Many passerines have streaked backs formed by the aligned, dark centers of the back feathers. On some species, the pale edges of the back feathers align to form lines known as *braces*. *Mantle* is sometimes used as an alternate term for the back, but it is better to reserve this term to describe the back and the scapulars together; mantle is rarely used in describing songbirds.

■ **Rump** The *rump* is located between the back and the uppertail coverts; the folded wings often conceal these feathers *(see page 50)*. This area is visible on flying birds, but on a perched bird it is uncovered only when the wing tips are drooped below the level of its tail. The rump feathers are loosely arranged and can be fluffed up to cover much of the wings during cold weather or, for some species, in a form of display. The rump can be distinctly colored and important for identification, as is the case on Cliff Swallow and Yellow-rumped Warbler.

■ **Uppertail Coverts** The uppertail coverts are neatly arranged rows of feathers that cover the base of the tail. There may be multiple rows, but the bottom row is the most prominent; there is one covert feather associated with each tail feather. The tail feathers and tail coverts move as a group when a bird moves its tail.

The Underparts The feathers of the underparts blend into each other, so the regions described by the various terms introduced below are approximate. For instance, there is no hard line where the breast stops and the sides begin, but knowing their approximate locations will get you looking in the right places. The following list of terms proceeds from the bill to the tail. The photograph of Dickcissel *(see page 61)* has all the following terms labeled.

■ **Chin** The *chin* comprises the tiny feathers directly below the bill.

■ **Throat** The *throat* is located below the chin and framed by the malar feathers on either side. The throat feathers move with the head as it turns, and sometimes there is a visible line separating them from the breast feathers below. Many passerines have useful field marks on the throat. Some species are even named for their throat color: Red-throated Pipit, Bluethroat, Yellow-throated Warbler, and Black-throated Sparrow, for example.

■ **Breast** This group of feathers occupies the shape of a downward-pointing triangle directly below the throat, and it blends into the sides and the belly. The breast feathers often have a pattern: Streaks and spots are very common. Sometimes a *central breast spot* (or "stickpin") coalesces at the lowest point of the breast. You'll find this distinctive feature on American Tree Sparrow.

■ **Sides** These feathers are located on either side of the breast and above the flanks. On birds with streaked underparts, the pattern found on the sides is often slightly bolder or otherwise subtly different from the other underparts. Where they border the wings, the side feathers are often fluffed up and conceal the lesser coverts and most of the median coverts.

■ **Flanks** The somewhat elongated and broader feathers of the *flanks* cover the body area below the wings. Streaking or other patterns that begin on the sides often extend onto the flanks. The flank feathers can be fluffed up to cover the lower part of the wing, but this is most common on waterbirds, such as ducks, loons, and grebes.

■ **Belly** The *belly* feathers are located below the flanks and breast and extend down to the undertail coverts. Much of the skin in this region is unfeathered, and feathers from surrounding areas loosely fill it out. The brood patch—skin that becomes highly vascularized to incubate eggs—is located in this region.

■ **Undertail Coverts** The *undertail coverts* are a group of feathers organized in overlapping rows that cover the underside of the tail, similar in arrangement to the uppertail coverts. The undertail coverts are also known as the *crissum*—as in Crissal Thrasher, which has distinctive chestnut-colored undertail coverts. The *vent* is the region where the belly feathers meet the undertail coverts. Vent (or cloaca) is also used to describe the single opening through which birds pass both reproductive and waste products.

The color, length, and pattern of the undertail coverts can be important to identification. For example, the color of the undertail coverts is usually whitish on Tennessee Warbler, compared to yellow on the related Orange-crowned Warbler. It is sometimes helpful to notice how far the tail extends beyond the longest undertail coverts:

On Tennessee Warbler the exposed tail beyond the undertail coverts is short; Orange-crowned Warbler has a much longer, exposed tail. This same distinction quickly separates the longer-tailed Pine Warbler from the shorter-tailed Bay-breasted and Blackpoll Warblers *(see photographs, page 98)*. Patterns, especially streaks, are also common on the undertail coverts.

The Legs and Feet The structure and proportions of the legs and feet are quite variable in birds. Leg length varies from very long to very short; feet have adapted to perching and swimming and have even become tools for tasks as varied as killing prey and digging nest holes. They have evolved to meet the specific demands of each species' lifestyle and feeding niche. For identification purposes, the most important features to consider are usually the color and length of the legs.

■ **Legs** The bones in a bird's lower leg correspond to the bones of a human foot, and a bird's foot is actually composed of toe bones: Birds walk on their toes. None of that particularly matters for identification purposes, although it explains why a bird's "knee" seems to bend the wrong way: It's an ankle. On birds, the section of leg directly above

This breeding male **Dickcissel** has all the main feather groups of the underparts labeled. *(Ohio, May)*

the foot is called the *tarsus* (plural, *tarsi*). On passerines, the tarsus is sticklike, covered with scales (or *scutes*) on most species, and it has very little muscle tissue. The leg section above that is called the *tibia*, and it is hidden by the flank feathers on most passerines; this section is well muscled and corresponds to the drumstick of a cooked chicken.

tibia feathering —

tibia —

tibio-tarsal joint — or "knee"

tarsus —

toes

The tibia is partially or fully feathered. The joint between the two sections is called the *tibio-tarsal joint* and is known vernacularly as the *"knee."* A third section of leg connects to the tibia—the real knee and the thigh (femur bone)—and is rarely visible on any birds.

As already noted, leg color can be an important field mark. In some species leg color can vary at different times of the year or as a bird matures: For instance, the yellow legs and feet of a spring male Blackpoll Warbler become darker by fall, when only the soles of the feet remain distinctly yellow. In general, leg color is a more important field mark (and is more visible) on nonpasserines, especially egrets, shorebirds, and gulls.

Leg length does not vary much within the passerines. Species that forage on the ground, such as jays or meadowlarks, tend to have longer and heavier legs than those found on arboreal or aerial-feeding species, such as flycatchers or swallows. Among the nonpasserines there is much greater variation: Consider the vastly different leg lengths of pelicans and cranes, two species of similar

A shorebird, not a passerine, the aptly named **Black-necked Stilt** rises above its surroundings on exceedingly long, coral pink legs. Passerines, on the other hand, have short legs, and the "knee" and tibia are often completely hidden by the belly feathers. *(Florida, June)*

bulk. On large flying birds it is often possible to assess leg length by noting how far the feet and legs project beyond the tail. Most birds fly with their legs trailing behind them, but take into consideration that some species can hide their legs and feet beneath their belly and flank feathers.

■ **Feet** The structure of a bird's foot varies considerably, even down to the number of toes. Passerine feet are all rather similar—each species has three forward-facing toes and one hind toe—and there are few usable field marks to take note of. Within the nonpasserine families, feet show considerably more variation: Webbed, partially webbed, and lobed toes are adapted to swimming; extremely long toes facilitate walking on floating vegetation; large talons grip prey; and woodpecker toes are adapted to clinging to vertical surfaces. Even so, feet are rarely useful as field marks because closely related species usually show the same adaptive structure.

In the remaining sections of this chapter we will delve into some of the specialized feather groups and bare-parts terminology of the non-passerine families. The next section starts with waterfowl and then continues on to tubenosed seabirds, herons, raptors, shorebirds, gulls, and hummingbirds—following the most recent taxonomic sequence.

Waterfowl

Waterfowl are an ancient, worldwide order of web-footed, gregarious birds that frequent aquatic habitats. In North America, all 62 species of waterfowl that have been recorded are placed in one large family—the ducks, geese, and swans (family Anatidae). The geese and swans

The violet-blue speculum is highly visible on this adult male **Mallard** in flight. When it lands, the wing folds and the scapulars, tertials, and flank feathers hide the speculum and most of the other wing feathers. *(Ontario, April)*

A male **Mallard** floats high in the water, its wings hidden from view by the scapulars and a big tertial feather from above, and by the flank feathers from below. *(California, January)*

are relatively homogenous groups. Ducks are more varied and are often subdivided into three subgroups: perching ducks, dabbling ducks, and diving ducks. The diving ducks are further subdivided into bay ducks, sea ducks, mergansers, and stiff-tailed ducks.

The Wings and How They Fold Even a casual observer can see that ducks don't look much like songbirds. Waterfowl may have the same basic body features and feather groups as songbirds, but the body proportions and feather shapes are very different. The feather locations on a swimming duck are often puzzling to beginning birders: The wings seem to disappear when a duck lands. Where did they go? The answer lies with the organization and length of certain feathers and how the wings fold.

In the photograph of the flying Mallard *(see page 63)*, the feathers of the wing appear to be organized much as they are on a songbird—primaries, secondaries, and coverts in neat rows. The black-and-white tips of the greater coverts that form a wing bar in the center of the wing are rather short and leave a lot of the secondaries exposed. The prominent secondaries on ducks often display a patch of iridescent color known as the *speculum*; this is violet-blue on the male Mallard. The tertials are silvery brown and extend past the trailing edge of the secondaries. The male Mallard's tail is white with two black central tail feathers that curl up decoratively. Notice the pale crescent above the orange legs that is formed by the rearmost flank feathers. The flank feathers will play an important role as the wing folds.

This adult male **Black Scoter**, of the distinctive North American subspecies *americana*, has a prominent yellow-orange knob at the base of the bill. On the underwing, the dark wing linings contrast with the silvery inner webs of the primaries. On the upperwing, the dark outer webs of the primaries can also be seen. *(New Jersey, February)*

taller head shape
with slight crest

black
restricted
to nail

coarser
vermiculations

rounder
head shape

larger nail
with more
black

whiter flanks with
finer vermiculations

The flight feathers slide over and under each other as the wing folds. The primaries slide under the secondaries and secondary coverts; the secondaries slide under the tertials that start to bulge rearward; then the whole wing slides under the scapulars. Upon landing, the lower edge of the folded wing is quickly tucked behind the flank feathers and hidden from view. On the swimming Mallard *(see page 63)*, the only wing feathers now visible are the big gray-and-brown tertials, the tip of the primaries, and a few white tips of the greater coverts. The grayish scapulars are in front of the tertials, and the two groups tend to blend into each other where they meet. This feather arrangement gives the bird a sleek shape that helps to protect it from the elements.

Sometimes just after landing or when actively diving, the wings are not tucked behind the flank feathers and are much more visible. This can give a very different pattern to the sides and be confusing when viewed from long range. Male ducks (known as drakes) often have highly ornamental tertials and scapulars with elongated shapes and fancy patterns and colors; female ducks (known as hens) tend to have smaller tertials and scapulars that often leave a bit more of the wing exposed.

The Bill The color and shape of the bill are often important field marks. Given that the swans have similar, all-white plumages, the bill color, bill shape, and the shape of the feathering at the base of the bill offer the best clues to their identification. Overall bill size helps to separate the smaller-billed female Blue-winged Teal from the larger-billed female Cinnamon Teal, but such discernment requires experience to appreciate the sizes, unless you can compare the birds directly.

Waterfowl have a hardened tip to the upper mandible, referred to as the *nail*. Sometimes the size and color of the nail are useful field marks as well. The photographs of the two scaup, Lesser and Greater *(above)*, show how similar these two species can be. The larger, more

The two species of scaup are very similar bay ducks in the genus *Aythya*. The black on the bill tip of **Lesser Scaup** *(left)* is limited to the small nail; on **Greater Scaup** *(right)* the black bleeds beyond the larger nail. These two birds are adult males, but the difference of their head gloss (purple or green) is not a reliable field mark. Head shape is useful on birds that are not actively diving: Greater Scaup has a rounded head; Lesser Scaup has a taller and narrower head that often shows a slight indentation (notch) on the rear of the crown. Lesser Scaup's bill tends to have straighter sides, and Greater Scaup tends to have whiter flanks and finer vermiculations (thin wavy lines) on the scapulars and flanks. Try to find several field marks when making a difficult identification. *(left: California, January; right: Ohio, February)*

The plumage of **Northern Fulmar** varies from light to dark; this is an intermediate bird typical of many West Coast birds. Although they superficially resemble gulls, these tubenoses are stocky birds with heavy bills, and they fly with stiff, outstretched wings. Notice the shadow lines outlining the scapulars from above and below. (Alaska, May)

extensively black nail that is found on Greater Scaup is a small but useful difference.

Tubenoses

The tubenoses—albatrosses, shearwaters and petrels, and storm-petrels—are pelagic seabirds that spend most of their lives far off-shore and are rarely seen by "land lubbers." You're going to have to get on a boat to see these species with any regularity; organized pelagic birding trips are available on both coasts and have expert leaders aboard. Some species are very similar to each other and extremely difficult to identify. Although chumming with fish refuse or popcorn is used to lure birds close to the boat, it is much more typical to see birds in flight and from a distance. Flight patterns of the wings and tail are very important, as are details of the bill—if you can get close enough to observe them. With experience, flight style is another important clue to identification.

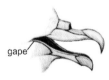

These two drawings show the typical bill of a *Pterodroma* or gadfly petrel. The bill plates have named parts that apply to all the tubenoses, from albatrosses to storm-petrels. The *naricorn* is the tube that encloses the nostrils. The strongly hooked *nail* (or *maxillary unguis*) is adapted to grasp slippery prey.

The Bare Parts Tubenose bills are different: They are made up of individual plates joined by fleshy seams and have *nostril tubes* located on the top—or on albatrosses, the sides—of the bill. Nostril tubes are prominent on some species, as shown in the photograph of Northern Fulmar above, but more important identification clues are usually the color, pattern, and overall size of the bill.

Leg length can also be useful for identification. The long legs of Wilson's Storm-Petrel extend well past its tail in direct flight, in contrast to other North American storm-petrels with similar plumage. Flying tubenoses can—and often do—tuck their legs forward, hiding them completely with their belly and flank feathers and obscuring any

The pale edges outlining every feather on this **Greater Shearwater** delineate all the major feather groups. There are about 20 secondary feathers contributing to the long-winged aspect. The humerals are visible, but they are not nearly as extensive as on albatrosses. (New Jersey, July)

scapulars

uppertail coverts

rump

tail

nasal tube

secondaries

lesser coverts

greater coverts

median coverts

bend of the wing (carpal joint)

alula

median primary coverts

greater primary coverts

primaries

inner secondary coverts

humerals and humeral coverts

scapulars

tail

humerals

secondaries

feet

primaries

This endangered **Short-tailed Albatross** is in adult plumage and has a wingspan of over seven feet. The dark patch formed by the humerals and greater humeral coverts is clearly visible. The white scapulars overlay the humerals like a cape and smooth the transition between the wing and the body. Note how the feet project beyond the tail; the two other North Pacific albatrosses also have short tails. (Torishima, Izu Islands, Japan, October)

differences in leg length. Sometimes leg or foot color is important, but it can be difficult to observe. All tubenoses have webbed feet; the yellow color of the webs on Wilson's Storm-Petrel *(below)* is distinctive when seen.

The Wings Tubenoses are often observed in flight, so it is important to know the locations of the various feather groups of the wings. The flying Short-tailed Albatross and Greater Shearwater are typical of the long-winged groups, and the upperwing feather groups are labeled on both photographs *(page 67)*. The wing shape—long and narrow with a pointed wing tip—is the most efficient for rapid sustained flight with frequent gliding and relatively few wingbeats. Albatrosses have extensive *humerals* and *humeral coverts* that contribute to the length of their wings. The humeral feathers, a third set of flight feathers and coverts associated with the uppermost wing bone (humerus), overlap the secondaries. An albatross folds its extremely long wings in three parts: The humerals fold back, the secondaries forward, and the primaries back. Only in this way can the wing be folded into a compact shape. The humeral feather group is absent or concealed on species with shorter wings.

The pale *carpal bar* on the wings of many storm-petrels slants diagonally across the inner wing and is formed by various rows of secondary coverts. As shown in the photographs *(below)*, pale carpal bars are very noticeable on Leach's Storm-Petrel and less so on Wilson's Storm-Petrel. The *M-pattern* found on various shearwaters and petrels is formed by dark primaries that connect to dark carpal bars that meet at the rump.

Wilson's Storm-Petrel *(left)* and **Leach's Storm-Petrel** *(right)* are both white-rumped species and are superficially similar, but they differ in many features and are placed in different genera. The long, swept-back wings and forked tail of Leach's are completely different from Wilson's shorter, rounder wings and squared-off tail. Leach's flies with deep wingbeats and its bounding flight often changes direction; Wilson's flight is usually fluttery. Wilson's has long legs and long toes with yellow webs that it patters on the sea surface while foraging. *(left: New Jersey, July; right: Atlantic Ocean, April)*

shorter wings with more rounded shape

very long legs

yellow webs

uppertail coverts divided by dark line

prominent carpal bar

forked tail

long pointed wings, held swept back

crest of
filoplumes

gular pouch

Breeding **Double-crested Cormorants** engage in a variety of ritualized displays. The male *(top)* presents nest material and shows off his brightly colored gular pouch in a display known as kink-throating. In the western subspecies *(albociliatus)* the color of the skin and eyes is somewhat variable, and the color of the crest feathers varies from white to mostly black. *(California, April)*

Pelicans and Allies

The pelicans are part of a large assemblage of waterbirds in the order Pelecaniformes that includes tropicbirds, boobies and gannets, pelicans, cormorants, darters, and frigatebirds. All the members of this order have *totipalmate* feet—feet with webs connecting all four toes— although the webs are greatly reduced in the aerial frigatebirds. The foot color of the boobies varies between species and is useful for identification, but it is complicated by its variability between different populations and ages. For instance, young Brown Boobies can have dull pinkish feet (yellow on adults), similar in color to the feet of young Red-footed Boobies (bright coral red on adults).

Gular pouches—throat pouches of bare skin—are present in all pelecaniform families except the tropicbirds and are particularly evident on frigatebirds, pelicans, and cormorants. The gular pouch serves various

These two **Snowy Egrets** show how radically different the color of the facial skin can be: The bird with crimson lores is in high breeding condition. Adult Snowy Egrets have a crest year-round, but they grow more and longer crest feathers before the breeding season. The crest can be raised in a threat display. (both: Florida, March)

functions: It can expand to hold large fish, be fluttered to regulate body temperature, and be used for display. In the gular display of some species, called kink-throating, the pouch is bulged outward by the hyoid bone located in the neck. The color of the gular pouch and the facial skin intensifies prior to courtship, as does the color of the eyes and even the skin inside the mouth.

Anhingas and some cormorant species molt in white filoplumes—filamentous feathers present on other birds but rarely visible—in various locations such as the thighs, neck, and back as the breeding season approaches. The prominent white thigh patches on Great, Pelagic, and Red-faced Cormorants in breeding plumage are composed of these filamentous feathers.

Herons and Egrets

Most species in this elegant family of wading birds have long legs, long necks, and long bills for stalking food in shallow water. Herons have dark plumage, while egrets are largely white, but the distinction is artificial and not based on taxonomy.

Long, graceful plumes called *aigrettes* grow from the scapular region and adorn egrets in breeding season. Other wispy plumes on the crest, the lower neck, and the back are visible year-round on some species, but these feathers are often replaced or augmented by longer feathers before courtship begins. Commercial plume hunters in the pursuit of aigrettes decimated the populations of Great, Snowy, and Reddish Egrets around the turn of the 20th century, when these feathers were in great demand for ladies' fashion. Concerned women in Boston organized in opposition to the slaughter of these birds, and their efforts led to some of the United States' first conservation laws—and the rise of the Audubon movement.

In *high breeding* condition, the bare parts of herons and egrets develop intense colors. The bare facial skin of the lores is often a striking color, and the legs and feet change color on some species. Coloration is most vivid during the prebreeding courtship period but can change from moment to moment.

Raptors

The hawks, kites, and eagles are a large, worldwide family of *diurnal* (active by day) birds of prey with hooked bills, strong talons, and highly developed flight abilities. The falcons and caracaras are a related family of fast-flying raptors with long wings that are bent back at the wrist and, except on Crested Caracara, narrow and pointed.

The Head and Bill Many features of a raptor's head are helpful for identification. Check for patterns around the eye and bill, looking especially at the facial stripes on various falcons and the presence or absence

This **Great Egret** is in high breeding condition and is displaying its long, lacy aigrettes that grow from the scapular region. This species' facial skin is yellowish most of the year but becomes vivid lime-green during courtship. *(Texas, March)*

of a pale supercilium. The size and color of the bill and the color of the iris vary among many species and are easiest to observe on perched birds. The *cere* is the fleshy covering at the base of the upper mandible that surrounds the nostrils. The color of the cere and the orbital ring are useful for determining the age of some species. On large falcons, these areas are bluish gray on juveniles and yellow on adults.

The Wings Flying raptors have distinctive wing shapes that help to identify them. Important areas to study are the shape of the wing tip, the length and width of the wing, and the shape of the trailing edge. The narrowing of the outer primaries on hawks and eagles breaks up the wing tip into individual *fingers,* which improves soaring and slow-speed flight capabilities. The slots between the feathers highlight the relative length of the individual feathers: For instance, a soaring Broad-winged Hawk presents a more pointed wing tip because it has a shorter outermost primary and fewer fingers than Red-shouldered Hawk, which has a squarer wing tip and more fingers. Wing shape changes with different flight activities; soaring flight is the easiest to study and gives the most consistent shapes for making comparisons.

Red-tailed Hawk is our most common buteo—a widespread genus of high-soaring hawks. This adult eastern Red-tailed *(borealis)* has the features of the head and wings labeled. An adult Red-tailed is also shown in flight opposite. *(Connecticut, February)*

diffuse supercilium

yellow cere

nape

scapulars

lesser coverts

breast

median coverts

greater coverts

belly band

tertials

alula

secondaries

primaries

tail

wing tip shorter than tail tip

dark comma

patagial bar

subterminal tail band

fingers

streaking forms diffuse belly band

This adult eastern **Red-tailed Hawk** (borealis) is soaring, and all the features of the under-wings are visible. The pale rufous tail and the dark patagial bar on its forewing make this an easy identification. (New Jersey, October)

translucent crescent-shaped window

Some buteos, like this juvenile eastern **Red-shouldered Hawk** (lineatus), can be tough to identify. The translucent crescent-shaped windows on the wing tips are distinctive but hard to see. Experienced hawk watchers would notice the square-tipped wing shape, different from that of Broad-winged Hawk, a species with similar juvenal plumage. (Pennsylvania, October)

wing coverts

scapulars

American Kestrel is a small colorful falcon: Its pointed wings and long tail are typical of all falcons. This adult male has blue-gray wing coverts and a row of translucent white spots on the trailing edge of the wing. Females have rufous wing coverts and a finely barred tail, and they lack the row of wing spots. (New Jersey, September)

On perched birds, take note of both the distance from the wing tip to the tail tip and the *primary projection*—how far the wing tip extends past the tertials.

With few exceptions, the plumage of the underwings holds more clues to a raptor's identification than the plumage of its upperwings does. Look at the extent of barring, the distribution of light and dark areas, and the color and pattern of the wing linings. A few areas have specific terms. The *patagium* is the leading edge of the inner wing; a dark *patagial bar* is found on most Red-tailed Hawks. A *window* is a pale translucent area at the base of the primaries; you can see crescent-shaped windows on a soaring Red-shouldered Hawk. A *carpal patch* is a dark area formed by the primary coverts and located near the bend of the wing, or wrist.

The Tail Most species have distinctive patterns of light and dark bands or bars on the tail. On soaring birds the tail is usually spread, and patterns are visible from both above and below. It is often easier to assess tail shape when the tail is closed: The spread tails on soaring birds all look rounded or fan-shaped. A *terminal band* is a band located on the tip of the tail, and a *subterminal band* is a band just in from the tip.

The Body On flying birds, especially soaring buteos, the feather patterns on the body hold few clues. Look for any color or pattern on the undertail coverts. Streaking on the underparts is common, and sometimes the streaks or spots coalesce into a dark *belly band*. There is more time to study the underparts of perched birds, and subtle variations in patterns can be discerned. When considering the upperparts, pay particular attention to any pattern on the scapulars: Red-tailed Hawks have pale mottling there.

Spotted Sandpiper *(left)* and **Semipalmated Plover** *(right)* are in different families—the sandpipers and the plovers—but they and a few others are all referred to as shorebirds. These birds are both juveniles. *(New York, August)*

Color morphs are variations in plumage color that occur regionally or within the same population. Different color morphs are common in some raptor species. Intermediate and dark-morph birds have darker bodies than light-morph birds but the same tail patterns. Intermediate morphs usually show underwings with at least a trace of the light-morph pattern.

Shorebirds

North American shorebirds represent seven families and about 90 species, but many are regionally uncommon, and some are rare to accidental. Birders that actively look for shorebirds can expect to see 30 to 40 species a year. These birds have a reputation for being difficult to identify: Many species are small- to medium-size birds with plumages of muted browns and grays, and they are typically seen in mixed flocks on distant mudflats. Separating them sounds tough, but an experienced birder is often able to identify many of the local species at a glance by concentrating on size, shape, and behavior (known as *jizz*). Shorebirds are quite vocal, and their calls are another important clue. If you're just starting out, get familiar with the

crown
supercilium
back
brace (formed by back feathers)
upper scapulars (3 rows)
lower scapulars (2 rows)
tertials
tail
bend of wing hidden by side of breast
lesser coverts
median coverts
greater coverts
wing tip

features for the most common species in your area: Spotted Sandpiper and the Semipalmated Plover *(opposite)* are common in many areas at the right time of year, and each has a distinctive jizz. Once you are familiar with the common species around you, you will know to slow down and take a closer look when something with a different shape or behavior stands out—just like the experts do.

When you take a closer look at a bird that strikes you as different, you need to know what to look for to clinch an identification. It might be a structural detail like bill or leg length, but often you need to know plumage details or even a suite of field marks. Shorebirds have the same body plan as other birds, but certain feather groups are especially important. If you come away from this section on shorebirds with an understanding of their scapulars and tertials—where to look for them and how they form patterns—you will be ahead of the curve and on your way to working out even the trickiest identifications.

All the important feather groups of a standing shorebird can be seen on this juvenile **Pectoral Sandpiper**. The feathers look very neat and fresh because they were grown a few short weeks before, when the bird hatched in the Arctic. By contrast, fall adults often look ratty because their feathers are old and worn. *(Ohio, September)*

The Body The upperparts of standing shorebirds are dominated by the extensive scapulars and the coverts and tertials of the folded wings located below them. You can follow along by referring to the photograph of the Pectoral Sandpiper *(above)*. The back feathers (or *mantle*) are small and occupy a small triangular area bounded by the scapulars on either side and behind the head. These three feather groups—back, scapulars, and wings—tend to blend into each other on a standing bird: Look for shadow lines and contrasts in feather sizes to separate them. All the feathers generally get larger and longer toward the rear and from top to bottom. The rump and uppertail coverts are mostly hidden on standing birds.

The *scapulars* grow like a graduated bunch of flowers from the small area where the wing attaches to the body, so the individual scapular feathers nearest the head are short and those visible at the rear (near the tertials) are much longer. The scapulars are organized into horizontal, overlapping rows. The *upper scapulars* (or shoulder scapulars) consist of three rows of smaller feathers, often with a different pattern. Below them are two rows of *lower scapulars,* larger feathers that contrast in size with the smaller wing coverts below them. In comparison to juveniles, adults tend to have larger and longer lower scapulars that cover more of the wing coverts. On some shorebirds, such as Pectoral Sandpiper *(previous page),* the white edges of the scapulars and the back feathers align to form pale lines known as *braces.* If these pale lines connect at the back of the bird, they are said to form a *mantle-V* (back feathers) or a *scapular-V.* The visibility of the scapulars can change with a bird's posture or state of alertness: The relaxed feathers on the Pectoral in the photograph drape over the wing coverts; at other times the scapulars are sleeked back and expose more of the wing coverts, as seen on Semipalmated Sandpiper *(opposite).*

The underparts have their share of important patterns—for instance, the streaked breast that ends abruptly on Pectoral Sandpiper—but there are no features that are unique to shorebirds.

This juvenile **Red Knot** has raised its scapulars to preen. Try locating the division between the lower scapulars and the wing coverts. The complete pale fringes on the feathers give the plumage a scaly look. The dark subterminal fringes just in from the pale fringes are diagnostic of juvenile Red Knot. *(Ohio, September)*

fluffed up scapulars

tertials

wing tip (tail hidden)

The partial webbing between the toes—semi-palmation—is visible in this image of a juvenile **Semipalmated Sandpiper**. This alert bird has sleeked down feathers. The wing is exposed as if the bird is ready to fly, and the exposure of the scapulars is minimized. (*New Hampshire, August*)

The Wings Working with a standing bird (consider again the photograph of Pectoral Sandpiper), it is best to work forward from the rear end. Look for the dark tips of the projecting primaries. The *primary projection* past the longest tertial is an important feature on many species. If you can locate the tail—the primary tips often hide the tail—this is another good reference point. The long wings that extend past the tip of the tail on White-rumped and Baird's Sandpipers help to separate them from similar shorebirds. Just forward from the folded primaries are the long *tertials*; three or four are usually visible. The tertials often have patterns, and those patterns are important field marks on some species. The secondaries are usually completely hidden beneath the tertials.

In front of the tertials and below the scapulars are three or more curving rows of wing coverts. The first row, next to the tertials, is the *greater coverts,* the next row is the *median coverts,* and then there are multiple rows of *lesser coverts,* similar to the arrangement on all birds. The lower row of scapulars overhangs the wing coverts. The *inner greater coverts* are the large prominent feathers adjacent to the tertials; importantly, they often show the same pattern as the tertials.

On flying birds *(see Least Sandpiper, page 78),* some species show contrasting patterns on the upper wing. *Wing stripes,* usually formed by the tips of the greater coverts and the pale bases of the primaries, are common and vary in length and intensity between different species. Unlike on standing birds, the rump and tail are visible in

On flying shorebirds, you can check for wing stripes and note the pattern of the rump and uppertail coverts. Like most peeps (small sandpipers in the genus *Calidris*), this juvenile **Least Sandpiper** has a moderate wing stripe and a dark rump pattern that extends onto the uppertail coverts and central tail feathers. The scapulars are tightly folded and cover most of the tertial feathers. (California, August)

flight and their patterns can be important. The flying Least Sandpiper pictured above shows a typical pattern found on many small shorebirds, especially in North America: dark rump, dark central tail coverts, and dark central tail feathers. This makes the white-rumped shorebirds conspicuously different in flight—especially both yellowlegs and both dowitchers, but others as well. Underwing patterns and the color of the wing linings vary and can be useful on species like curlews and godwits, and it is easier to get a look at the underwings of these large species.

The Head and Bill The feather groups and their names are the same as for the other groups of birds already discussed. A *split supercilium* is a small extension of the pale supercilium that starts near the bill and extends into the dark crown feathers above the main supercilium.

Bill length and shape are highly variable in the shorebirds, and thus they are important to identification. A good yardstick is to compare the length of the bill to the length of the head. Some bills are obvious and striking; the long, down-curved bill of Long-billed Curlew comes to mind. Assessing bill length can be problematical on some species: Females have longer bills than males, so a long-billed female may have a bill as long as a short-billed male of a different but closely related species. Many bills are all-dark, but check for any color at the base of the bill.

The Legs and Feet As with bills, leg length is extremely variable between many species. On flying birds with long legs, the feet and sometimes part of the legs extend past the tail. Noticing the amount of extension—if there is any—can be useful. Leg color can be obscured when birds feed in muddy areas; the yellow legs of Least Sandpiper often end up looking blackish for this reason. Partial webbing between the toes is known as *semipalmation,* and this attribute gives its name to both Semipalmated Sandpiper and Semipalmated Plover.

Gulls

The gulls and terns form two distinctive subdivisions of the family Laridae, and the skimmers complete the family. There are 49 species of gulls and terns recorded from North America—29 gulls, 20 terns, and one skimmer. The numbers, large as they are, belie the bewildering array of plumages available for study. The larger gulls that take about four years to attain adult plumage—known as *large white-headed gulls*—have intervening plumages that are varied and variable. Many of the small gulls have dark hoods as breeding adults—and are thus referred to as *hooded gulls*—and mature in just over one or two years. Your field guide will help you sort them out, but it is helpful to know some gen-

This adult **Herring Gull** (smithsonianus, top) shows the features found on the head of a typical, large white-headed gull species. The petite, ternlike **Bonaparte's Gull** (bottom) is one of the hooded gulls that has a dark slaty hood as a breeding adult. The Bonaparte's shown is in juvenal plumage and is probably two or three months old. Ear spots are typical of the smaller hooded gulls in immature or winter adult plumage. (top: Ohio, February; bottom: Ohio, August)

eral feather topography and terminology specific to gulls. Gulls are a conspicuous part of our North American birdlife.

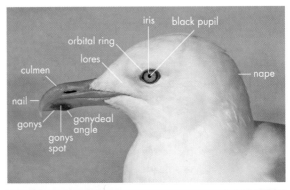

When observing gulls, try to estimate a bird's overall length and bulk before you move on to the feather details. You might try mentally sorting a gull flock into small (Bonaparte's size), medium (Ring-billed's size), and large (Herring Gull's size and larger) birds or into dark-mantled and light-mantled birds. Males are larger than females. If you have a large flock of medium-size, light-mantled Ring-billeds in front of you, this approach will help you find something different. You don't want to be spending all your time checking every Ring-billed.

The term *cycle* when used in gull terminology—first-cycle, second-cycle, and so on—refers to the cycle of complete feather replacement; each cycle can take about a year in gulls, sometimes longer. These terms are gaining acceptance as the preferred terminology for describing gulls because they relate to the plumage you see.

The Head and Bill Much gull study centers on the head and bill; we've provided two examples *(see page 79)*. Most or some of the head feathers molt twice a year, and those molts produce two different head patterns. On adult hooded gulls the patterns are very different. Large white-headed adults show less seasonal variation; most have some dark head streaking in winter. Always check eye color, judging its paleness or darkness. The color of the fleshy *orbital ring* is easiest to see on breeding adults and is often a confirming detail on some of the larger species. Critical study of the size, pattern, and color of the bill is important. Size and bill structure are fairly constant, but color and pattern change with age on many species. To judge the shape of the bill, consider the depth of the *gonydeal angle* (the angle on the lower margin of the bill as it slopes up to the bill tip) and the shape of the *culmen* (the shape of the bill's ridge line).

Standing Gulls Standing gulls often allow close approach and prolonged views. After you've had a look at the head and bill, assess the color and pattern of the back, scapulars, wing coverts, and tertials. Then shift your attention to the rear and observe the primaries for any patterns, their overall length, and their relationship to the tail; be sure to also check leg color. The following discussion will follow that sequence and refer to the photographs of the near-adult and juvenile Herring Gull *(below and page 83)*.

The term *mantle* is used often in gull discussions—but with differing meanings. Here, we use mantle as an all-inclusive term describing the feathers of the back, scapulars, wing coverts, and tertials

This third- or fourth-cycle **Herring Gull** is adultlike, but it still shows some traces of immaturity: blackish subterminal marks on the tertials, very limited tiny white apical spots on the primary tips, and dark marks on the bill. Note one white mirror visible on the underside of p10 on the far wing. *(Ohio, January)*

visible on a standing gull. On adult gulls these feathers are all the same color (from light gray to blackish gray), and those colors are consistent within most species. Thus, the terms *dark-mantled gull* (black-ish gray mantle) and *light-mantled gull* (all the rest) separate the various species into two, easier-to-manage groups; a few species are intermediate.

The term *back* is reserved for the feathers behind the nape and between the scapulars on either side. As on shore-birds, on gulls the scapulars are promi-nent and hold useful field marks, particularly in the patterns on imma-ture birds. Adults of some species have a white patch at the rear of the *scapulars* known as a *scapular crescent*. To the rear of the scapulars are the tertials. These large feathers often stack up to form an abrupt break where they bor-der the primaries, known as a *tertial step*.

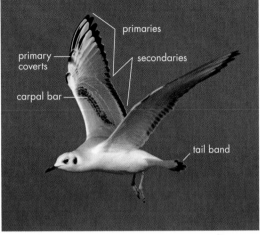

This first-cycle **Bonaparte's Gull** looks much like an adult on the body, but the tail and wings have patterns absent on adults, as described in the text. By the following winter, it will be in adult plumage. (Ohio, December)

On adults, the tertials are often tipped with white and form a *tertial crescent*. Sometimes the white tips of the secondaries connect to the tertial crescent and form a white lower border to the wing, known as a *skirt*. In front of the tertials are the rows of wing coverts: a single row of large *greater coverts,* a single row of smaller *median coverts,* and mul-tiple rows of much smaller *lesser coverts.* The forward edge of the wing is often tucked slightly behind the feathers of the sides and flanks.

Shift your attention to the rear of the bird and focus on what you can see of the folded primaries and tail. Gulls are long-winged birds, so you will normally see many folded primaries behind the tertials. Look for any patterns and check the color (light to dark). Part of the tail can usually be seen too, so you can check the *primary projection past the tail* by estimating how far the wing tip extends past the tail. Sometimes the undersurface of the wing on the other side of the bird is visible; when this is the case, you can check for any patterns there.

The underparts offer useful clues for identifying immature gulls; adult gulls in North America are primarily white below, except for Heer-mann's Gull. The color of the legs and feet is definitely worth exam-ining. Leg color is variable, but doing a rough mental sort by color—pink, yellow, and black—might lead you to something different.

Flying Gulls When watching a gull flock, standard procedure is to start with standing birds, using a scope if it's available, to get an idea of what's out there. Gull flocks eventually get restless or spooked and often lift off in a swirling mass. As this begins, you can try to stay with a bird that looked interesting on the ground and follow it with your binoculars as it takes off. Overall size differences can be significant between species, and this can be more apparent in flight.

On flying gulls, first decide whether the bird is an adult or an immature. This identification is important because you'll want to look at different parts of the bird depending on its age. Most adults have clean white underparts, uniformly light gray or dark gray upper wings with black-and-white wing tips, and white tails; most immature birds have mottled bodies, browner or boldly patterned upper wings, and all-dark or banded tails. In temperate regions of North America, birders are always on the lookout for *white-winged gulls,* northern species that have paler wings at all ages, such as Glaucous and Iceland Gulls.

Starting with the immatures, refer to a small and a large species: a first-cycle Bonaparte's Gull *(previous)* and a first-cycle Herring Gull *(opposite)*. Bonaparte's Gull is a small gull that matures in two plumage cycles, so, as with others of this type, much of the upperparts are already the gray color of the adult. The upper wings, however, show clear signs of immaturity: a dark carpal bar, blackish marks in the primary coverts, more extensive black on the primaries, black secondaries, and a black tail band. A carpal bar is a contrasting diagonal bar across the coverts of the inner wing. The underwings of the small hooded gulls can show distinctive patterns; for instance, adult Little Gull has dark gray underwings that are diagnostic.

Herring Gull takes much longer to reach adult plumage—four plumage cycles—and the progression of change is slower. First-cycle Herrings are variable brownish birds with subtle but definite patterns. You need to know what to look for on the various species; consult your field guide. On first-cycle Herrings, though, the best mark is the paler inner primaries that contrast with surrounding flight feathers. Additional marks to observe on other species include any patterns on the tail, dark or light areas on the inner wing, and the rump and uppertail coverts.

For an example of an adult in flight, we'll stay with the widespread Herring Gull *(above)*. Remember that on distant birds you'll want to ask yourself some preliminary questions: Is this an adult or an

This adult **Herring Gull** shows the head streaking present in winter; otherwise it is similar to individuals in breeding plumage. A large mirror is located on p10 and a small one on p9. Three tongues of gray are located on p6, p7, and p8. *(Ohio, January)*

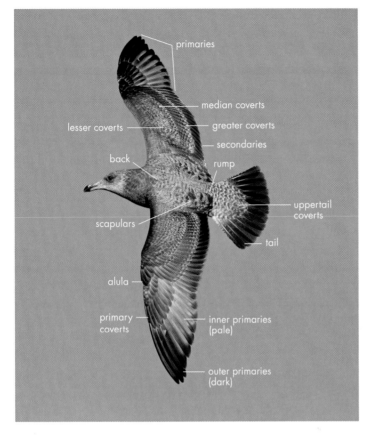

A first-cycle **Herring Gull** (smithsonianus) is a study in brown; the challenge is to find something to focus on. Assess the overall size and bulk of the bird, and move on to looking for any especially dark or light areas. The inner primaries are noticeably pale on first-cycle Herring Gulls. If you see an all-brownish gull in midwinter, you can be sure it's one of the larger four-cycle species. In contrast, all ages of Ring-billed Gull—a three-cycle species—have light gray backs and scapulars by midwinter. (California, November)

immature? Is it small, medium, or large in size? Is it light-mantled, dark-mantled, or possibly white-winged? Often you can make a quick, almost snap decision on these questions. For our Herring Gull example, the answers would be, "Adult, large, and light-mantled." Sometimes the combination will get you most of the way or all the way to a species identification. In the East, if you exclude rarities, those answers get you to Herring Gull. The closest contender—don't forget that identification is a problem on distant birds—is the medium-size Ring-billed Gull. In the West, in addition to considering Ring-billed, you have to consider California and possibly Thayer's or a hybrid between Western and Glaucous-winged.

There are some special features on the wings of adult gulls that can help with separating birds, so you should be familiar with their terminology; they are labeled on the photographs. A *mirror* is a small white spot just in from the tip of an outer primary. Check to see whether your gull has more than one mirror, and note the mirror's size and location (the primary on which it occurs). The mirror can be seen on the underwing, but patterns there are usually muted. Assess the amount of black on the wing tip and the extent that the gray penetrates the black, known as *tongues*. The primaries usually have white tips, called *apical spots,* and these vary in size. The white trailing edge of the inner primaries and secondaries is very prominent on some species. If the bird flies close by, you might be able to see bill, eye, and foot details.

Hummingbirds

Hummingbirds are one of the largest New World families of birds, with over 300 species recognized. North of Mexico the number dwindles to just 23 species, of which 14 breed regularly. In eastern North America there is only one breeding species, the Ruby-throated Hummingbird, but some western species wander to the East in fall and winter. Birders in western North America have many more species to become familiar with on an everyday basis. The adult males of most North American species are relatively easy to identify, but females and immatures present identification challenges that may require a careful look at small details. Out-of-range hummingbirds always require close inspection to verify their identity.

Observation With their tiny size and rapid, insectlike flight, hummingbirds might appear to be impossible subjects to study, but experienced birders know that with patience, many details can be observed.

This adult male **Ruby-throated Hummingbird** is easily identifiable in this view. Note the contrasting black chin that extends in a band back under the eye; this black band is lacking on other "red-throated" hummingbirds (Broad-tailed and Anna's). Sometimes it helps to notice the small details: The contrast between the narrow inner primaries and the broad outer primaries separate Ruby-throated from all other North American species except Black-chinned, the only other member of the genus *Archilochus*. Note the very slender tip to the outermost primary (p10); this feather is broad and blunt-tipped on Black-chinned. *(New Brunswick, June)*

black chin

secondaries

narrow inner primaries (p1–p6)

broad outer primaries (p7–p10)

p10 with slender tip

tail

Hummingbirds are very territorial and often return to a favored perch, allowing close inspection and even views through a scope. This is your chance to look at wing details, such as the proportions and shape of the primaries, as well as head and bill details. Birds hovering at feeders or other nectar sources present the best opportunities to study tail patterns: While the wings are a blur, the rest of the bird can appear remarkably stationary. It helps that most species are almost fearless and often allow a close approach, seeming to trust in their remarkable flight abilities to cope with any danger. Even so, many hummingbirds do not stay in view long enough for a confident identification to be made, and some distinctions—between immature Rufous and Allen's, for instance—are usually impossible in the field. When starting out, concentrate on becoming familiar with the more distinctive adult males; then move on to the females and immatures.

p10
p9
p8
p7
p6
p5
p4
p3 p2 p1

ten primaries
(p1–p10)

six small
secondaries

r5
r4
r3
r1 r2

The ten tail feathers are numbered r1 to r5 on each side: the two central tail feathers are both r1; the outermost tail feathers are both r5.

Separating an immature or female Rufous Hummingbird from the nearly identical Allen's remains one of our most difficult field identifications, and most birds are best left as Rufous/Allen's. But in a good stop-action photograph—such as this one of an immature male **Rufous Hummingbird**—details of the shape and pattern of the tail feathers are much easier to study and are fairly reliable for identification. The relatively broad outer tail feathers (r3–r5) and the nipple-like tip to r2 on this bird indicate Rufous. By midwinter many immature male Rufous Hummingbirds have mostly rufous backs, unlike Allen's. *(California, August)*

Structure The structure of a hummingbird appears quite different from other species of birds, but hummingbirds have the same basic body plan of other birds. The parts differ only in their proportions. The two photographs—one of a perched Ruby-throated Hummingbird *(opposite)* and one of a Rufous Hummingbird in hovering flight *(above)*—have labels pointing out important features. The prominence of the primaries and the short, reduced secondaries are easily seen on the perched Ruby-throated, whereas tail details are visible on the hovering immature male Rufous.

Iridescence Hummingbirds are noted for their metallic iridescence. These brilliant colors are produced by light reflecting off the internal structure of each feather, not by pigments. As such, these colors change with the light or the viewing angle, and sometimes they appear to turn on and off like a light bulb. The *gorgets* (throat feathers) of adult males are especially variable, appearing to be flat black in some situations. The iridescent green feathers on the upperparts of most species have a different structure that reflects light over a broader angle than that on the gorget feathers, so the upperparts color is less variable and usually always visible.

lores
crown
postocular spot
ear coverts
nape
gorget
lesser and median coverts
scapulars
breast
greater coverts
primary coverts
secondaries
flanks
inner primaries
foot
outer primaries
vent
band
uppertail coverts
wing tip
tail

This illustration of an adult male **Rufous Hummingbird** is life size. The major feather tracts visible on a perched hummingbird are labeled.

CHAPTER 5

HOW TO
IDENTIFY BIRDS

M any beginning birders find bird identification a challenging subject, fraught with complications and pitfalls. Hundreds of species and thousands of plumages are illustrated in most North American field guides. There are so many variables. How can one possibly make sense out of all that information?

After you've had a few experiences birding with others or even alone, you'll discover that it's not all that difficult. Many birds are actually easy to identify, and if you take into consideration what is likely in your location and the time of year (status and distribution) the field is substantially narrowed. Most species are distinctive in some or even many ways—like the White-throated Sparrow at left. Humans are a visual species, with brains geared to latch onto the distinctive patterns, colors, and behaviors that their eyes take in. You may not have memorized a bird's name yet, but you can train yourself to remember what you saw (or heard) long enough to discuss it with a fellow birder or flip through the proper section of your field guide. Over time, you will develop your own internal field guide.

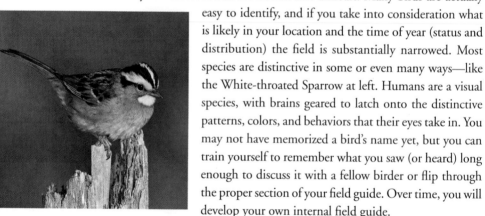

Black-and-white head stripes, a bright yellow supraloral spot, and a white throat make this adult **White-throated Sparrow** (above) distinctive. After a few sightings you just know it. A flock of waterfowl (opposite) is more challenging. This image contains numerous **Greater Scaup**, three **Common Goldeneyes**, two **Redheads**, a **Canada Goose**, and an **American Coot**. (above: Michigan, May; opposite: New York, February)

When you first start birding, you will see birds that confuse you or that you don't see well enough to identify. Just let them go, for the time being. As a beginner, you have so many new birds to see that sometimes it can be better to move on to something else.

As your birding experience and skills grow—and this happens early on—you'll want to give some thought to the process of identification and ways of approaching some of those difficult-to-identify species. This does *not* mean memorizing your field guide. It means knowing enough about bird identification to know how to look at a bird and how to process what you see. This chapter will present some of the variables you'll need to consider: size, structure, plumage, and behavior. Each section starts with an easy or even obvious example and proceeds to more subtle and challenging ones.

An Approach to Identification

In recent years, some authors and birders have made a distinction between different "schools" of bird identification: the traditional field-marks school versus the general-impression-of-size-and-shape school, the latter of which is also known as the GISS (or *jizz*) school or as birding by impression. We make no such distinction. Rather, we think that using aspects of both approaches is the best course. Depending on the viewing situation and the species under consideration, everything you see is grist for the identification mill.

After you have accumulated some field experience, you'll identify many everyday birds at a glance, relying on your impression of the species. It's a wonderful way to sort through a flock; you don't want to spend all your time examining the same common and familiar birds. Something different and more interesting might be lurking there.

In other situations—often involving species you aren't familiar with—it might be crucial to notice, for example, that the scapulars and greater coverts are marked with anchor-shaped crescents or that the wings extend past the tail. These feather details can be important; sometimes they offer conclusive evidence that is lacking in impression birding. That's why a basic understanding of bird topography should be part of your birding knowledge. You need to know your way around the various feather groups of a bird to even notice some of these details.

Seeing a suite of field marks—*multiple* distinctive features that *together* are unique to a species—is the best approach to getting the right identification. A common pitfall for beginners is latching onto a single, obvious field mark. For example, in summer, the adult male Indigo Bunting and Blue Grosbeak are both blue birds (but they are not bluebirds—those are thrushes). You cannot rely on the blue color to identify the males of either of those species. In addition to plumage field marks, Blue Grosbeaks often pump and partially spread their tails, especially when landing (jizz details) and they give loud, explosive *chink* calls (voice). These features, taken together, add up to breeding adult male Blue Grosbeak, without question. Seeing just one of these features leaves the identification inconclusive. Consider that your mind can play tricks on you. If you want to see a Blue Grosbeak, it's only human nature to fill in the blanks of that blue flash with the features you are hoping to see. Take a good look and make sure they are really there.

Judging Size

Size is a rather vague term when used alone. What we're really referring to is *overall* size. If you want quantitative numbers, you can use the numbers published in every field guide—usually the average length (bill tip to tail tip) of a series of museum specimens. Of course, nobody is out there measuring birds in the field (except bird banders), but those published numbers are not a bad starting point. You can compare the average lengths of various species and usually reach the proper conclusion about which species are larger and which are smaller. The *relative* overall size, not the exact length in inches or centimeters, is the important distinction.

Judging size in the field can be tricky and has led to many misidentifications. Even with something for comparison—a nearby bird is best—accurate size assessment is best viewed as approximate. The most extreme field situation is a featureless sky or ocean; there is no way to tell how far away a bird is even if you can compare it to others.

In summer, adult male **Indigo Bunting** *(top)* and **Blue Grosbeak** *(bottom)* are both superbly blue, but notice the subtle difference between the blues. You need something more than that single field mark—blue—to make the identification. In addition to its color, Blue Grosbeak's suite of field marks includes (in order of importance) chestnut wing bars, a heavier bill, larger size, and a more extensive black mask, in addition to its different voice and behavior. *(both: Ohio, May)*

In addition, birding optics (binoculars, scopes, and cameras) distort the relative sizes of birds. This effect, which makes distant birds appear larger than similar-size birds that are closer to you, is known as *size illusion*. You can test this illusion yourself by looking at a rectangular object such as a picnic table through your binoculars. Get in a position parallel to the front edge and take a look through your binoculars: The table will look trapezoidal, with the far edge appearing wider than the near edge! Or place two same-size coins on a table with one coin 12 inches closer to you, stand back, and take a look: The coin that is farther away appears larger. Size illusion, of course, also distorts our perception of birds in similar situations.

Keeping these caveats in mind, size can sometimes be a useful field mark. Take your time and try to observe the bird in question from various angles and distances. The safest approach is to have other known species for comparison, and never use size as your only field mark.

Mixed Species Groups Comparing overall sizes can be particularly useful when observing flocks of mixed species in the same field of view—waterfowl and shorebirds come to mind. Snow Goose and Ross's Goose or Greater Yellowlegs and Lesser Yellowlegs, when seen together, are very different in overall size, and this can be easy to see. When they are seen alone, the identification becomes more difficult.

Subtle differences in overall size can also be useful. Imagine you're looking at a group of small calidrid shorebirds (birders refer to them as peeps) on an East Coast mudflat. At first glance they all look alike, but as you study them, you start to see some size differences—perhaps your first clue that more than one species is present. With further study you conclude that the smallest birds are Least Sandpipers. First you notice they look brownish rather than grayish and tend to feed closer to vegetation rather than on the exposed mud; then you check bill shape and plumage details. Those a bit larger are Semipalmated Sandpipers. And the largest one in the group—it has longer wings than the others—could be White-rumped Sandpiper, but this requires further study. Overall size differences were your first clues, and then you confirmed your identifications by looking at plumage details. In many cases, overall size won't get you further than, "That brown bird is larger than the sparrows I've been seeing. I think it's something different. Maybe it's a thrush." But you've already narrowed the field and have a starting point.

Single Birds Many species are rarely or never seen in flocks, however. If you see a bird by itself, you might get useful clues from something

else in its field of view—such as the size of a leaf, a flower, a bird feeder—but you'll have to use your memory to compare it to another species. Hairy and Downy Woodpeckers are good examples. These widespread species overlap in range and have nearly identical plumage. They are very different in size but are rarely seen next to each other. Because of this, size differences are not especially useful. You'll probably need something more to seal the identification *(see caption on page 92)*. Beyond the structural differences labeled in the photographs, a key plumage field mark is whether the outer tail feathers are pure white (Hairy) or have dark bars (Downy). This is often difficult to observe, and dark subspecies of Hairy in maritime Canada and the Pacific Northwest can have tail barring. Calls are diagnostic once you've learned them. Here's another example of how a suite of field marks locks in the identification.

Structure

The term *structure* refers to the parts of a bird and their proportions. Many structural features are related to the underlying skeletal framework of the bird. Field guides customarily point out a species' long legs, short bill, big head, and small eyes—all structural terms. The relative difference in bill size between Hairy and Downy Woodpeckers discussed above is a structural difference. Birders also refer to the wings and tail as structural features: a species might be called long-winged or short-tailed, variations that depend on a combination of feather length and skeletal features.

Adult **Ross's Goose** and **Snow Goose** are virtually identical in color and pattern, but Ross's is smaller and weighs about half as much as Snow Goose. There are also some structural differences to look for: Note the smaller bill, rounder head shape, and shorter neck of the Ross's. On standing birds you could study bill details: Snow Goose shows a blackish "grinning patch" on the bill, and the feathering at the base of the larger bill is more indented. *(California, February)*

Downy Woodpecker *(left)* and **Hairy Woodpecker** *(right)* are virtually identical in color and pattern. Above they are reproduced with the same body size to draw attention to the *proportional* differences of their bills. The same photographs appear below, but here the birds are reproduced in the proper size relationship. *(both: Minnesota, February)*

Shape and Feathers

The *shape* of a bird is another vague term that needs some clarification. Although the underlying structure and proportions are very similar among all individuals of a particular species or subspecies, the actual shapes that are visible can vary. Feathers are the reason for this: They can be held tight to the body or raised (fluffed), often depending on a bird's activity *(opposite, top)*. Head and body feathers are the most affected. Roosting birds often fluff their feathers for warmth, giving them a larger more rounded body; active birds appear more slender by comparison. Excitement level—whether it is related to courtship, feeding, or other activities—may affect specific feathers, like the raised neck feathers on a displaying prairie-chicken or the raised crest on a Greater Roadrunner. Individual birds, even in the same situation, can have different shapes, sometimes for no apparent reason. For example, a birder might relate the experience of seeing a large Great Horned Owl and a skinny Long-eared Owl roosting next to each other. However, normally, the Great Horned would have the Long-eared for dinner, so that's a very unusual friendship. A more likely scenario is that two Long-eared

Owls were roosting together (not unusual) and that one had adopted a camouflage pose, squeezing all the feathers tightly together in an attempt to look like tree bark, whereas the other bird was unconcerned and relaxed, allowing its feathers to expand, and so appeared twice the size of the other.

With all this variability in shape, what is useful? What is objective? The bare parts—bill, eyes, legs, and feet—are mostly unfeathered, so they should be unaffected, right? No. In fact, the feathers around bare parts can affect one's assessment of a bird's size. Take a look at the two photographs of Semipalmated Sandpiper: One bird is fluffed-out and roosting, and the other is in an alert, upright posture. The alert bird looks longer legged because the tibia (upper leg) is fully exposed, and it appears somewhat longer billed because the head looks smaller. Our brains automatically assess the size of the parts in relation to overall size, and the fluffed up bird looks bigger and heavier, which makes the exposed parts, like the legs and bill, look smaller. The take-home lesson is that there are limitations to using shape and proportion as field marks because our perception of them is so affected by how a bird is holding its feathers.

What Works Some structural details are so obviously different that they always help with identification. For example, the long, down-curved bill of a Long-billed Curlew is radically different from the slightly upturned one of a Marbled Godwit, even though their size and plumage are rather similar. More subtly, the flight silhouette of a soaring Turkey Vulture (with its long, narrow wings and long tail) is always different from that of a soaring Black Vulture (which has wide wings and a short tail).

These are examples of useful structural field marks, and they are nearly always visible. Even distant birds can be told apart after a few experiences with them. You will find many others examples mentioned in your field guide. The following photographic examples, presented in taxonomic order, illustrate some different techniques for assessing useful structural field marks.

■ **Eared and Horned Grebes** Eared Grebe is primarily a western species, but because it can show up just about anywhere (it's a good

Although these juvenile **Semipalmated Sandpipers** (probably the same individual) show very different shapes and sizes, under all those feathers they are identical. With experience, you'll learn what is (and what is not) reliable as a structural field mark. Differences in shape are particularly prone to variation and misinterpretation. (both: New Hampshire, August)

peaked crown with peak above eye

steep forehead

thinner, slightly upturned bill

longer, thinner neck

fluffy rear (variable)

relatively flat crown

sloped forehead

thicker neck

tapered rear

slightly thicker, straight bill with white tip

In winter, **Eared Grebe** *(top)* and **Horned Grebe** *(bottom)* can be separated by a combination of structural and plumage details. The labels point out structural and shape differences; plumage differences are mentioned in the text. *(top: California, January; bottom: California, February)*

find in the East), you cannot make an identification based on range. Eared and Horned Grebes are similar in size and have similar plumage patterns in winter, so we will first cover plumage differences and then move on to the structural features. Horned has a clean white lower face, with the white cut off abruptly by blackish gray at the level of the eye; a white lores spot is usually evident in front of the eye. Eared has a dingier face, with gray blending into white well below the level of the eye, giving it a dark cheek. Overall body shape is highly variable, but Eared Grebes often ride higher in the water and have a fluffed-up rear. This broad characterization of overall body shape might get you to look more closely at a distant bird, but it is not a definitive field mark. Structural differences of the head and neck are much more useful: Eared has a steep forehead with a peak near the front, whereas Horned has a more sloped forehead and a flatter top of the head. Eastern's slightly thinner bill, which often shows a noticeable upturn near the end, differs from Horned's thicker and straighter bill, which has a small whitish tip. Eared also often shows a thinner and longer neck than Horned does, although posture can cancel out these differences.

■ **Red-tailed and Swainson's Hawks** Flying raptors are a challenging group to identify. Many books have been written on raptor identification: We recommend the classic *Hawks in Flight,* published in 1988, which emphasizes structure and behavior. Our example just scratches the surface of this subject, but it may encourage you to look for more than plumage field marks when hawk watching.

Red-tailed Hawk is the default hawk across most of North America, and if you want to get good at hawk identification you should take every opportunity to look closely at this species. It is the baseline to which all other species are compared, so you need to know it well. Our example compares a juvenile Red-tailed to a juvenile Swainson's Hawk. There are important plumage field marks to see, such as Red-tailed's dark patagial bar and the Swainson's much darker flight feathers; but the structural details are also important and useful from a greater distance or if the light is poor. Start out with an overall assessment of the bird's impression, its jizz: Red-tailed is a bulky raptor, and the blocky wings and the bulging secondaries gives it a pumped-up, muscular look; Swainson's is altogether more graceful and slender, and the tapering wings and long tail give it the lines of a runner or dancer. This assessment is pretty subjective, so let's back it up with some hard facts. Focus on the structure of the wing tips. On Red-tailed, the four longest primaries are of similar length and give the wing tip a blunt, slightly rounded shape; those same primaries on Swainson's Hawk form a pointed wing tip. Always confirm your impressions with these kinds of hard facts when you can, and remember to take into consideration what type of flight you're observing. These photographs show soaring birds: The best opportunity to observe wing details is when birds have their wings fully outstretched to generate maximum lift. A gliding bird pulls its wings in with the wing tip swept back, a position that allows it to speed up, usually while losing altitude.

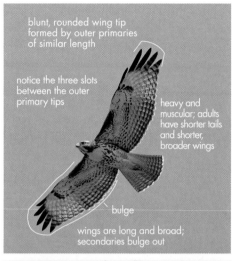

blunt, rounded wing tip formed by outer primaries of similar length

notice the three slots between the outer primary tips

heavy and muscular; adults have shorter tails and shorter, broader wings

bulge

wings are long and broad; secondaries bulge out

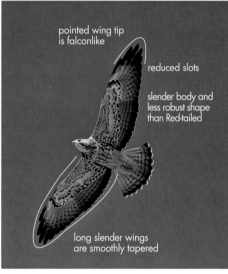

pointed wing tip is falconlike

reduced slots

slender body and less robust shape than Red-tailed

long slender wings are smoothly tapered

■ **Black-bellied Plover and American Golden-Plover** These two species, along with Pacific Golden-Plover and European Golden-Plover (a casual spring migrant to eastern Canada), make up the genus *Pluvialis*. Black-bellied Plover is much more common than any of the golden-plovers. As with Red-tailed Hawk, Black-bellied Plover is the default when considering any of the golden-plovers. If you think you have a golden-plover, start off by asking yourself, Why isn't it a Black-bellied?

Let's compare juvenile Black-bellied Plover to juvenile American Golden-Plover (see page 97). With a name like golden-plover, you might think that golden coloration would be a good clue. Although our photographs seem to show that, the reverse can also be true: Some juvenile Black-bellieds have a strong golden cast. A few plumage features are worth noting. For instance, while juvenile Black-bellied has a sharply streaked breast, American Golden has gray-brown barring that gives a quilted look. And the white supercilium of juvenile American Golden is prominent *behind* the eye, set off by a dark cap above,

Compare soaring juvenile **Red-tailed Hawk** *(top)* and juvenile **Swainson's Hawk** *(bottom).* Flight identification of raptors is an art and a science. Experienced hawk watchers rely on "soft" field marks (impressions of shape and movement) to assess distant birds, but they also know the "hard" field marks to look for. *(top: Arizona, August; bottom: Colorado, September)*

and more dark in the auriculars, whereas Black-bellied has a fainter supercilium with internal streaks. On some birds these plumage features are less than obvious, and size and structure are a better guide.

The overall size and shape (an impression) can be difficult to judge depending on the bird's activity, but in this case the general perception is that Black-bellied is a bulky bird with a larger more robust bill; American Golden gets called dovelike and has a smaller, rounder head and a dainty bill. A close look at the rear of the bird can give you something concrete to assess—the lengths of the exposed tertials, primary tips, and tail. Look at the annotations on the photographs *(opposite)* and also notice the long primary extension on American Golden: The wing tip extends well past both the tertials and the tail. The overall look is of a long, tapered (attenuated) rear half. If you're close enough, you can count the white-edged primary tips. The other golden-plover that breeds in North America, Pacific Golden-Plover, has very long tertials that cover more of its primaries: Usually three tightly spaced primary tips are visible. Both golden-plovers breed in the Arctic and undertake marathon migrations that use their long wings to the fullest. American Golden-Plover winters as far south as Argentina; Pacific Golden-Plover winters in south Asia, on islands in the tropical Pacific, and as far south as Australia.

■ **Pine and Blackpoll Warblers** Small passerines are often very similar in structure, so the differences among them tend to be difficult to see, unless you are specifically looking for them. Adult and immature male Pine Warblers and fall Blackpoll Warblers, both in the genus *Dendroica,* have similar plumages in fall *(see page 98).* There are numerous plumage differences between them, but in this example we will focus on the length of the tail that extends past the longest undertail coverts—an unambiguous difference. The third member of this look-alike trio, Bay-breasted Warbler, has a short tail extension like Blackpoll, a close relative. In fall, those two species must be separated on plumage details. Poorly seen birds are often left as "Baypoll Warblers." Adult Pine Warblers look similar year-round, but Blackpoll and Bay-breasted Warblers have very different, easy-to-identify spring plumages.

Plumage: Pattern and Color

Not only are the patterns and colors of a bird's feathers beautiful to see; they often hold the most important clues to the bird's identity. In this section we'll look at carefully chosen examples of plumage patterns and colors that are important to identification. The plumage of a bird, sometimes called its *feather coat,* is a highly organized system of different types of feathers that serve different purposes. We discussed the names and locations of the various feather groups in chapter 4 and some aspects of how they affect a bird's structure (long winged, short tailed, etc.) in an earlier section of this chapter. Next, we will discuss the patterns and colors of feathers and how to assess some of the classic field marks they produce.

Field Guides Many beginners ask, "Why don't they arrange field guides by color? That's what I see first and what's most important to me." In chapter 2 we touched on why all the best field guides are arranged taxonomically with related species shown together. Briefly, there are so many variations within each species and among closely related species—between males and females, between different ages, between different times of the year—that arranging them by color would result in chaos. Not only would closely related species often be found sections apart,

Juvenile **American Golden-Plover** *(top)* and **Black-bellied Plover** *(bottom)* have very similar plumage patterns. Relying on your impression of their different size and shape (jizz), can leave you guessing. More reliable are the consistent structural differences of the tertials, primaries, and tail. On difficult identifications, use all the field marks available to you. *(both: Ohio, October)*

long extension of the tail past the undertail coverts

short extension of the tail past the undertail coverts

Adult **Pine Warbler** *(left)*, whose plumage shows little seasonal variation, resembles the fall **Blackpoll Warbler** *(right)*. A structural difference that holds true year-round is the long extension of the tail past the undertail coverts in Pine Warbler versus the very short extension in Blackpoll Warbler. The dark auriculars, sharply contrasting with the paler chin, are diagnostic for Pine Warbler in all plumages. *(left: New Hampshire, April; right: California, August)*

but birds with equal amounts of different colors would also get arbitrarily assigned to one or another group, and males and females would often get separated.

True beginners usually know more about bird families than they give themselves credit for. Many groups are obvious to almost everyone—ducks, egrets, hawks, woodpeckers, and numerous others—and these groupings are all found together in your field guide. When you're starting out, a useful goal is to become familiar with the accepted taxonomic sequence of families and some of its representative members.

Pattern Human eyes and brains are exquisitely attuned to patterns, such as strong light and dark or color contrasts. Even newborn babies are fascinated by strong patterns. The black-and-white pattern of an adult Black-throated Sparrow *(opposite)* is immediately memorable and easy to identify. It helps that there are no other sparrows with similar patterns. As happens with many species, one plumage is very distinctive, whereas the plumage of other sexes, ages, or times of year are different *(see chapter 6)*. The juvenile Black-throated Sparrow lacks the adult's black throat patch, which results in a much less distinctive pattern.

Few species are quite as boldly patterned as an adult Black-throated Sparrow, but many plumage patterns are obvious and important, such as wing bars, head stripes, streaking, and barring, to name just a few. Through a combination of field experience and studying your field guide, you'll find examples of patterns on just about every species. Patterns can be as simple and distinctive as the black-and-orange of an adult male American Redstart or as intricate and subtle as the internal markings of the tertials and inner greater coverts on a juvenile Short-billed Dowitcher. (The very similar juvenile Long-billed Dowitcher has unmarked centers on those feathers.)

Always look for any patterns when you're birding; they're usually important for identification. Sometimes patterns are most useful at getting you to the right group of birds, not the exact species: Breeding loons, many woodpeckers, *Myiarchus* flycatchers, and chickadees come readily to mind. Once you've narrowed the field, an exact identification is much easier to make.

Your field guide will point out many patterns, but be open to discovering new ones. Something you discover for yourself is much more memorable than anything you read in a book. Common birds make good test cases, and sometimes field guides gloss over their identification. You might try looking at patterns on small birds in flight that visit your feeders. The white wing bar on the median coverts of a male House Sparrow, for instance, are prominent in flight but rarely mentioned in any field guide. Two examples of birds with pattern-type field marks are presented below.

■ **Western and Cassin's Kingbirds** These two yellow-bellied kingbirds are common in parts of the West and overlap in range. (Two other yellow-bellied kingbirds have much more limited ranges: Tropical Kingbird is local in southeast Arizona and the Rio Grande Valley of Texas, and Couch's Kingbird is common in the lower Rio Grande Valley.) A close look at tail pattern can often confirm an identification. Western Kingbird has a very dark, blackish tail and uppertail coverts, and the outer tail feathers have bright white outer edges that show as white stripes. The blackish uppertail coverts contrast with the pale gray back and rump. On Cassin's the tips of all the tail feathers are pale grayish white, and the outer tail feathers do not have bright white outer edges,

There's little to be said about the identification of an adult **Black-throated Sparrow** *(left)*. The black-and-white pattern of this desert sparrow is unique and simple to recognize. But there's a curveball: The juvenile *(right)* has a whitish throat! Unlike most sparrows, Black-throated Sparrow retains its juvenal plumage well into fall and migrates in this plumage. A first encounter has left many birders scrambling for a field guide, but notice the bold supercilium on both birds. *(left: California, November; right: Arizona, August)*

Western *(right)* and **Cassin's Kingbirds** *(opposite)* are similar flycatchers in the genus *Tyrannus* with overlapping ranges. The different colors and patterns on the head, chest, wings, and tail help to separate them. Voice is also an excellent clue: Western gives repeated sharp *kip* notes that are very different from the loud, burry *chi-bur* call of Cassin's. *(right: California, April; opposite: Arizona, May)*

darker face mask usually visible

pale chin blends into light gray breast

darker wings

bright white edge contrasts sharply with black tail

dark tip

although they may be pale there. The tail and uppertail coverts of Cassin's Kingbird are browner and not so dark; consequently, they show little contrast with its darker gray back and rump.

Conveniently, these tail patterns can usually be seen from both above and below. In flight, Western's pale back, blackish wings, and black tail with its white "racing stripes" make for a more "contrasty" bird compared to the rather uniform Cassin's, which also has browner wings. The pattern of the throat is another important field mark: Cassin's white chin contrasts with its dark gray breast, whereas Western's white chin blends into its pale gray breast.

■ **Cassin's and Botteri's Sparrows** These two sparrows, found in arid southwestern grasslands, are similar—very similar. You might consider this to be an example of the graduate-school level of birding. But even if you're a beginner in another part of the country, working through this tough identification can give you some perspective on the process of bird identification. Both species are members of the genus *Aimophila*: secretive sparrows with large bills and rounded tails. Becoming familiar with the similarities of sparrows in the same genus can help you become familiar with this large family of little brown birds. Taking note of the three photographs on page 102 is easier than making most field identifications, unless you have singing birds, since these species have very different songs. Your field guide will describe their songs, but it's better to listen to a tape or sound

darker gray
head with
no mask

pale chin
surrounded by gray

paler and
browner wings

off-white edges (sometimes
absent or worn off) show less
contrast than on Western

pale tip of tail

file (available online). We'll concentrate on the plumage details visible in the photographs on page 102.

Many birders make their first acquaintance with these two species during the summer months in southeastern Arizona. During those months, both these species and most other songbirds are looking a bit ragged. That's because their feathers are old and *worn* (abraded by the environment, especially any pale feather tips or edges) and bleached by the sun, which further obscures subtle patterns. *Molt,* the cyclical replacement of old feathers *(see page 128),* typically takes place after the breeding season but not during migration. Worn feathers will be replaced in late summer and fall. The differences of pattern and color between these two species will be most obvious during the winter months after the molt is complete. But by that time, Botteri's and most Cassin's Sparrows have migrated south into Mexico. The top photograph on the next page shows a Cassin's Sparrow in very worn plumage; the middle one is a Cassin's starting to molt and showing some fresh feathers and some worn ones; and the bottom one is a Botteri's that is worn, with no obviously new feathers visible.

A few plumage features of these birds are not visible on the photographs. Cassin's Sparrow has white tips on the outer tail feathers; Botteri's lacks this feature. The white in the tail is rarely visible on perched birds. But in flight it is usually evident, especially during Cassin's display flight, during which the sparrow flies up and then flutters down

two distinctive feathers with black subterminal crescents are visible

back feathers crossed with blackish bars

fresh tertial feathers are blackish with crisp white edges

faint wing bars

blurry flank streaks

uppertail coverts have black subterminal crescents (anchor marks)

arched culmen

back feathers have blackish streaks or stripes

no wing bars

no flank streaks

brown upper-tail coverts have black streak

to a different perch while singing. The pattern of the central tail feathers is also different: Cassin's gray-brown central tail feathers have a dark central shaft streak with jagged edges that gives it a serrated (or cross-barred) look; Botteri's central tail feathers are brown with paler brown edges. Both of these patterns virtually disappear on worn, summer birds.

This discussion of patterns addresses some very subtle distinctions. The example of Cassin's and Botteri's Sparrows relies on looking at small-scale features that are changed or obliterated by wear or molt. Most pattern-based field marks are not nearly so difficult to see; they're often bold and obvious. The next section will deal with plumage coloration—as opposed to pattern—and how it is used for identification.

Color Plumage colors can be startling in their intensity or cryptically intriguing in their complexity. Even the coal black of the American Crow has subtle variations of iridescence that change with the intensity and angle of the light. Many of us have (or will have) birding memories that revolve around color.

In partnership with pattern, color is integral to the way we see and identify birds. Colors are so central to the identification of birds that they are reflected in many bird names: "Blue This" and "Black That" or "Red-breasted This" and "White-tailed That." (*Common* appears to be the most frequent adjective in names. This is especially disconcerting when it is applied to a species that is only casual or accidental in North America, such as Common Crane or Common Sandpiper.) Some colors are so unique that they define our memory of a species—the cinnamon of a male Cinnamon Teal, the crimson of a male Vermilion Flycatcher, or the sky blue of a male Mountain Bluebird. Other colors come in unique combinations, like the steely blue and rust of a male American Kestrel, the metallic blues and greens of an adult Purple Gallinule, the chestnut cheek surrounded by yellow of a spring male Cape May Warbler, or the lime green upperparts and gray face of a fall Chestnut-sided Warbler.

■ **Color Descriptions** Problems sometimes arise with descriptions of certain colors as birders talk among themselves or in the written descriptions in field guides. One person's blackish brown is another's brownish black. As a starting point, decide what the overall color is, make that the noun (brown or black), and then modify it in whatever color direction you want: If the brown color leans toward black, it's blackish brown; if it leans toward red, it's reddish brown. When a color appears to be an equal mix of two colors, use both colors with a hyphen between them, as in blue-black or yellow-orange. A few color descriptions crop up frequently in birding and are rarely used in other situations: *Rufous* is the bright reddish orange-brown color found on the tail of Great Crested Flycatcher *(page 56)* or the slightly different color on Rufous Hummingbird; *glaucous* is the silvery, bluish gray color found on the back of some white-winged gulls, like Glaucous or Iceland Gulls; *roseate* describes any bird with a pinkish tinge on some part of its plumage, from the intense pink of Roseate Spoonbill to the pale pink cast of the underparts of a breeding Roseate Tern; and *flammulated* (literally, "in flames") refers to the reddish brown of the facial disk of Flammulated Owl.

■ **Color and Optics** It is important to remember that sometimes the colors of the plumage and bare parts, especially the paler colors, can be affected by birding optics. Most optics impart a colorcast to the image you see, usually making the image appear slightly warmer (yellower) or cooler (bluer). Check your own optics by looking at a large expanse of pale color through a single barrel of your binoculars, and alternately open and close your eyes. In most cases, as you switch back and forth between your binocular-aided eye and your unaided eye, you will notice a shift in color. And all optics drain some of the intensity from colors because the amount of light is reduced as it passes through the lenses and prisms.

■ **Color and Lighting** Our perception of colors is affected by the type of lighting present, but the brain automatically compensates for minor differences in lighting and alters our perceptions to moderate the effect. In the bright green light of a spring forest, the yellow of many warblers takes on a strong greenish cast, but the brain filters out the effect and sends the message that the color is clear yellow. If you were to look at an out-of-context photograph of the same yellow, you would clearly see the greenish cast. The same compensation takes place in

The photographs opposite were taken in Arizona in summer and show two **Cassin's Sparrows** (top and center) and a **Botteri's Sparrow** (bottom). When the birds are not vocalizing, the differences between these two species are subtle; identification relies on assessing small-scale differences of pattern on individual feathers.

The top Cassin's Sparrow is in very worn plumage, and any trace of its wing bars is gone. You might almost be able to identify it by its lack of pattern. Fortunately, two distinctive feathers are visible and positively identify this bird as a Cassin's Sparrow. Compare this bird to the Cassin's in the middle, which has fresher feathers showing clear-cut patterns. Newly molted feathers are on the back, and there is a single new tertial feather on each side.

The Botteri's is in worn plumage with no fresh feathers visible. A good structural detail to notice is Botteri's slightly larger bill with a curved culmen. Most Botteri's in Arizona (subspecies *arizonae*) are buffier on the underparts than this individual. *(top and center: Arizona, August; bottom: Arizona, July)*

the bluer light of early morning. Some light effects are so strong that they override the compensation. The warm light of late afternoon or sunset is especially potent and can affect our perception. Colors take on a strong orange tinge, sometimes obliterating color differences that are important to identification.

■ **Fading** The pigments that color feathers are subject to degradation by sunlight. Many species replace their feathers only once a year (species with different seasonal plumages molt some feathers twice a year), and by the time those year-old feathers are replaced, their colors may have substantially faded. Field guides often ignore this variation due to space constraints, but the effect can be dramatic. The black feathers of birds as different as storm-petrels and crows can look decidedly brown by the end of a year in the bright sun. On birds in molt, the new feathers coming in often are darker or more intensely colored; refer back to the Cassin's Sparrow photographs *(page 102)*, which show a worn and faded individual and another bird with new feathers growing in.

■ **Color of Bare Parts** The color on birds is not limited to feathers. Some identifications rely on seeing the color of the bare parts: bills, eyes, facial skin, legs, and feet. These colors can change with age or time of year in some species. This is particularly true in the larger non-passerines, many of which mature over a period of years. Some large gulls have distinctively colored skin around their eyes (orbital ring) that intensifies during breeding season, but the color only develops as the birds approach adulthood; younger birds lack the bright orbital ring colors. Bill color and pattern sometimes change as birds mature, although there are many fewer cases of this in the songbird families. A striking example of the change in color of the facial skin of Snowy Egret during courtship is shown in chapter 4 *(page 70)*. Many other examples exist.

Although we have discussed exceptions and complications to the use of color, it remains one of the cornerstones of the identification process. Colors and patterns are very important field marks on species that are similar in structure to other species; for instance, think of how many songbirds have similar structure. Three examples of identifications that rely on color and color comparisons are presented below.

■ **Female Summer and Scarlet Tanagers** Summer and Scarlet Tanagers broadly overlap in breeding range in both the Southeast and the

The lime green color of the cap and back of a fall **Chestnut-sided Warbler** is set off by its gray face, white eye ring, and pale gray underparts—a "gin and tonic with lime" combination that is unique in North America. Its wren-like posture with drooped wings and cocked tail is typical. *(New Hampshire, September)*

Midwest, with Scarlet breeding farther north and not nearly as far west. The all-red male Summer is more likely to get misidentified as Northern Cardinal (with a fleeting look) than to be confused with the red-and-black adult male Scarlet Tanager. The females are more of a problem. Paying attention to color can help out.

■ **Adult Ring-billed and California Gulls** Mantle color is an important field mark on adult gulls, and it is remarkably consistent. The Ring-billed Gull has a pale gray mantle color—almost identical to that of Herring Gull—and both are noticeably paler than California Gull *(see page 106)*. Sometimes it's easier to look at the contrast between the gray and white plumage to assess how dark or light the mantle is. Ring-billed's mantle almost blends into the white, whereas California's mantle makes a strong contrast. The bulkier California Gull also has distinct coloration on bare parts: It has dark eyes, grayish (or greenish) yellow legs, and, although there is a blackish ring on the bill, the pattern is different, and there is a red spot on the gonys. Ring-billed has pale yellow eyes, bright yellow legs, and a shorter, bright yellow bill with a wide black ring.

The angle of light and the background color can affect your perception. For any critical assessment, try to view the bird from various angles. Look at the two small photographs of the same California Gull on the next page. The bird against the white background appears to have a darker mantle. This visual effect can happen in the field, for instance, with a bird viewed against pale sand or a bright sky. Further, a bird that contrasts strongly with its background is usually interpreted as larger than one that blends into its background.

■ **Female Bluebirds** The identification of female bluebirds—Mountain, Western, and Eastern—is a study in subtleties. A quick glance won't do, and because these birds are variable, a suite of field marks is desirable.

The female **Summer Tanager** *(left)* and **Scarlet Tanager** *(right)* look quite different in color in these two photographs, but making that assessment when they are high up or poorly lit can be tougher. Look for the yellow color with mustard tones of the Summer versus the greener color with olive tones of the Scarlet. Female Scarlet Tanager also has a smaller and stubbier bill, rounder head, and darker wings. *(both: Texas, April)*

The Rosetta stone for separating female blue-birds is the color of the flanks and how that color contrasts with the undertail coverts *(see page 108)*: Female Mountain has grayish flanks (even when the breast is buffy) that contrast with white undertail coverts; female Eastern has bright cinnamon flanks that contrast strongly with bright white undertail coverts; and female Western has duller cinnamon flanks that contrast less strongly with grayish white undertail coverts.

Mountain Bluebird is the most distinct of the three: Structurally, it has longer wings that reach farther down the tail, a thinner bill, and longer legs, and it usually looks slender and graceful. Eastern and Western Bluebirds are more compact and similar in shape to each other, but Eastern has shorter wings than the other two. Female Mountain Bluebirds have a more concolorous (similarly colored, with little contrast) plumage: Notice how the buffy gray colors of the head, back, and breast blend into each other. On Eastern and Western females, the coloration of the upperparts is very different from that of the underparts. Many female Mountains lack warm tones on their plumage later in winter and are easier to sep-arate from female Easterns and Westerns, which always have rusty tones on the under-parts. In addition to the differences in flank

Compare the pale gray mantle of an adult **Ring-billed Gull** *(top)* with the darker mantle of **California Gull** *(center)*. Most see the same bird against the light background *(bottom)* as darker mantled and larger. *(top: California, November; center and bottom: California, January)*

color, Eastern has rusty coloration that extends onto the sides of the neck and wraps behind the gray ear coverts. Western's gray head con-trasts with its dull reddish gray back and dingy cinnamon breast. East-ern gives a musical, two-syllable *chur-lee* call that is quite different from the single mellow *tew* call of Mountain and Western. (Mountain's call is weaker and less musical.) Mountain Bluebird, the most migratory of the three, casually wanders to the Midwest and East in fall and winter.

Behavior

Behavior covers a wide variety of bird activities, many of which, though fascinating, are not directly applicable to birding and bird identifica-tion. Books on individual species or families of birds often delve deeply into this subject. *The Birds of North America* species accounts contain detailed life histories for every species that breeds in North America and Hawaii. (They are available in printed form, or you can access them online for a modest fee at *bna.birds.cornell.edu/BNA*.) The accounts in *The Birds of North America* are comprehensive ornithological works that were produced with the scientific community as their primary audience. Field guides lie at the other end of the spectrum: With so much information to cover and so little space, they offer scant details on behavior, doing so only when that information is crucial to identification. Thus a lot of interesting and useful behavioral information is unavailable in field guides. Some of what's missing in your field guide may be passed on to you by word of mouth from other birders or a mentor, and some may come by digging into the literature or even reading popular journals. The most rewarding learning will come from your own direct observations.

The classic side-view illustrations in field guides are static shadows compared to the experience of the living, breathing, moving bird. That side-view illustration packs a lot of information into a single image, but your experience of a bird is not going to be a close fit with your field guide's illustrations. Birds move, birds fly, birds behave—so your perspective on them is constantly changing—and no field guide can illustrate that experience. Therefore field guides are best thought of as shorthand visual reminders of the basic field marks and, if the artist is really talented, the illustrations will convey something about how a bird carries itself. Your experience of bird behavior in the field will be far richer and more detailed than anything you get from looking at illustrations or photographs.

Some of the most basic behavioral differences are often glossed over by field guide authors, who probably assume them to be known to everyone. You'll learn about these generic behaviors early on, but you'll learn much faster if you have a more experienced birder to clue you in. These birders will teach you basics like dabbling ducks rarely dive; plovers walk, stop, and pick but sandpipers are rarely so hesitant; terns plunge-dive and gulls don't; lots of species "fly-catch," not just flycatchers; vireos are sluggish whereas most warblers flit about nervously; and towhees jump-scratch in the leaves with both legs together. If you're a beginner, try to get out with others and ask lots of questions. Writing notes in a journal is a great way to learn about behavior; you'll remember your experiences much more vividly if you write them down. Later on, these early notes will be a great source of pride for you.

With time, your appreciation of the nuanced behavioral differences between closely related species will add another, very useful section to

The female **Mountain** *(top)*, **Western** *(center)*, and **Eastern Bluebirds** *(bottom)* vary subtly but consistently in color. Study the annotations and refer to the text for details on how to tell them apart. *(top: New Mexico, November; center: New Mexico, November; bottom: Texas, December)*

thin bill

no strong contrasts between head, back, or breast

long wings and short tail extension

gray flanks contrast with white undertail coverts

sky blue

gray head

no rusty color wrapping around ear coverts

dull reddish gray back contrasts with gray head and rusty breast

dingy cinnamon flanks contrast weakly with the grayish white undertail coverts

medium-length wings and tail extension

crown often bluish

reddish color of breast wraps behind gray ear coverts and contrasts with gray back

heavier bill

shorter wings and longer tail extension

bright cinnamon flanks sharply contrast with white belly and undertail coverts

your internal field guide. For some species, behavioral traits can be as important as structure and plumage when making an identification. The following examples cover a smattering of different types of behavior that will help you make a specific identification. Use them as samples of what to look for when making your own observations.

Flight For most people, flight is the defining characteristic of birds. Of course there are flightless birds, but none have been present in North America since the extinction of the Great Auk in 1844.

Inevitably, the study of flight styles and flight identification has tended to concentrate on large soaring birds like raptors and vultures, whose flight can be studied for extended periods of time. Books such as *Hawks in Flight* and the recent *Hawks from Every Angle* present detailed flight style information for most North American raptors *(see "Additional Reading")*. Hawk watchers distinguish between different types of flight such as soaring, gliding, kiting, hovering, and flapping. Different types of flight produce different silhouettes, and the visible plumage field marks will also be affected to some degree. Being familiar with the different silhouettes and which field marks can be seen from a distance (and which can't) allows the accomplished hawk watcher to identify distant birds with confidence.

In recent decades the popularity of organized pelagic (open ocean) birding trips has led to increased knowledge about the flight styles of tubenoses (albatrosses, petrels, shearwaters, and storm-petrels) and their importance in field identification. The frequency of wing beats and how they are interspersed with glides of various lengths is significant. For instance, on the West Coast, distant Pink-footed and Black-vented Shearwaters can look very similar in plumage pattern, and their different sizes are hard to judge at sea. The much larger Pink-footed has a lumbering flight style with slow wing beats (easy to count) and long glides; the small Black-vented has fast, flickering wing beats (too fast to count) interspersed with shorter glides. Wind speed must always be taken into consideration; at higher wind speeds, differences in flight styles are lessened because all species tend to flap less and glide more.

The identification of songbirds in flight is often challenging. Most songbirds are observed in flight only briefly, usually while they forage. Sometimes this foraging flight is distinctive: For example, kinglets tend to hover in front of vegetation while searching for insects; this is known as *hover gleaning*. Yet most flight behavior is not so obvious. Since most songbirds are nocturnal migrants, they are rarely observed in sustained flight. But even short flights and brief glimpses of the plumage patterns on flying birds can be informative. If you want to study the plumage of songbirds in flight, start out by training your binoculars on an active bird and continue until it flies. Very short flights are difficult to study, but if a bird moves to a different tree, attempt to keep it in view; limit yourself to describing one notable plumage feature or pattern and write it down. This simple exercise will train your eyes to see details you would otherwise ignore. Wing shapes vary and should be noted if possible. Most beginners soon learn the distinctive triangular wing shape of European Starling, but others are also distinctive.

Flight style is easier to observe without optics. Simply watch a known species until it flies, and then concentrate on the frequency of wing beats and glides (the rhythm): Note whether the flight line is straight or undulating, whether the wing beats are jerky or smooth, and whether the wings are open or closed during any glides. Group behavior can be useful: Some species are usually seen in flocks; others give distinctive flight calls. In this way you will get a feel for the gestalt of how the common birds in your area fly. High-speed photographs of songbirds in flight are beautiful to look at, but they rarely capture the effect of what you observe in the field. Songbird flight identification is not for everybody, but if you get interested you'll be making your own discoveries, since little has been written about it.

Spotted Sandpiper has one of the most distinctive flight styles of any North American bird. Its shallow, flickering wing beats, like vibrations, are interspersed with short glides on stiffly held, down-curved wings; the head is held high and a wing stripe is prominent. *(Ontario, April)*

Typical Movements The distinctive movements of many species are important to identification. Some of these movements are habitual and species specific: The tail-bobbing, teetering walk of Spotted Sandpiper is almost as unique as its flight style *(above)*. Whenever you watch a Spotted Sandpiper walk, this typical movement is seen. It does not depend on whether the bird is engaged in a specific activity. Other movements are coupled with specific activities, though, such as feeding or courtship displays.

The habitual movements that a species engages in can be extremely useful clues to identification. As with Spotted Sandpiper, tail bobbing is a consistent feature of the two waterthrushes, similar-looking warblers that spend most of the time walking on the ground. As you get familiar with these two species, you'll notice that there are consistent and usually quite noticeable differences in how they bob their tails; this is very useful for identifying which species you are looking at. Northern Waterthrush bobs its tail almost continuously in an up-and-down motion; Louisiana Waterthrush has a slower and more deliberate action that involves more of its body, and its tail moves in a side-to-side, somewhat circular direction.

Only five species of North American wood-warblers walk (rather than hop) on the ground: the two waterthrushes, Ovenbird, Swainson's Warbler (which shivers while it walks and regularly picks up leaves to forage from the undersides of them), and Connecticut Warbler. Except for the similar waterthrushes, these species are rarely confused with each other. The important point to remember is that Mourning and MacGillivray's Warblers—in the same genus as and often confused with Connecticut Warbler—don't walk; they hop. If you are hoping to see the uncommon Connecticut but your bird is hopping from branch to branch, keep looking. Connecticut Warblers walk on the ground and along logs or low branches with a deliberate gait; when flushed they often fly up to a low branch and sit motionless like a thrush.

The difference in silhouette and apparent size can be dramatic between active and resting birds. Sometimes field marks are obscured: The small notch on the hindcrown of **Lesser Scaup** (visible in the bottom photograph)—a very useful distinction from the rounder-headed Greater Scaup *(see page 65)*—is obscured by sleeked-down feathers after recent diving activity (as shown above). *(both: California, January)*

notch

Feeding Behavior Birds' feeding activities are often associated with distinctive movements, from simple, general behaviors (e.g., most woodpeckers cling vertically to tree trunks or branches as they feed) to species-specific behavior (e.g., American Dippers forage underwater, a unique behavior among North American songbirds). Somewhere in between the general and the unique is where most feeding and foraging behavior lies. A few more examples follow.

■ **Lesser Scaup** This example has a simple but important lesson: Activity can affect size and shape. When activity level increases, the feathers are normally compressed; the result is a smaller, more slender silhouette. Temperature has a similar effect: Fluffed-up feathers are better insulation in cold weather, and slimmed-down feathers trap less heat in hot weather. Waterbirds that get wet or dive during feeding—like the Lesser Scaup *(see page 111)*—are particularly prone to changes in shape because the water compacts the plumage during a dive. In diving birds, the air that is normally trapped by the feathers is squeezed out by water pressure, which reduces buoyancy and helps them dive.

Sometimes the difference in shape can obscure an important field mark. Plumage patterns can also change on actively diving birds: Some diving ducks and alcids—species that use their wings for underwater propulsion—often do not tuck their wings behind their flank feathers between dives. This can create confusing patterns, especially on birds viewed from a long distance.

■ **Reddish Egret, Snowy Egret, and Little Blue Heron** These three species can be all white: Little Blue Heron as a juvenile, Snowy Egret always, and Reddish Egret as a white morph. There are plumage and bare parts differences, but there are also differences in feeding behavior. The dashing Reddish Egret is the standout of the three; always in motion, it runs and pirouettes in shallow coastal waters with wings held high as it jabs left and right for small fish *(see page 2)*. Snowy Egret takes Reddish Egret's level of activity down a notch or two: Sometimes it's jerky and energetic like it just got hit by a jolt of electricity, sometimes it's almost motionless, and at other times it can be seen stirring the water in front of it by pumping one of its yellow feet on the bottom. Little Blue is a slow-motion specialist: When it moves, it goes slowly and methodically, rarely lifting its feet out of the water; when it stops—which is often—it assumes a neck-out, bill-down position, with its body tipped forward, and peers intently into the water.

■ **Stilt Sandpiper and Dowitchers** Stilt Sandpipers and dowitchers are often found feeding together in a similar manner. The larger yellowlegs resemble Stilt Sandpiper in coloration but feed by picking at the water's surface, not by probing. Stilts can look very similar to molting or basic-plumaged dowitchers: Birders typically see worn, breeding adults or molting Stilts with remnants of breeding-plumage barring on the flanks and belly, and juveniles in fresh plumage or molting into first-winter plumage (plain gray back and scapulars). Stilt's bill is long and dark, with a subtle, last-minute droop near the end—rarely the impressive tool wielded by dowitchers. If you flush a mixed group, look for Stilts' white rumps and mostly white tails, lacking the dowitchers' prominent white slash up the back. So while there are good plumage and structural differences, it is sometimes more straightforward to look at how the birds carry themselves and their feeding behavior.

Stilt Sandpipers forage almost exclusively while wading in the water, often belly deep. Because they have longer legs and shorter bills than dowitchers, they have to tip down more when they probe. As a result the tail is elevated when they probe; dowitchers tend to keep their bodies more horizontal. Stilts often submerse their heads completely and feed with a more deliberate probing than the mechanical stitching motion ("sewing machine") of dowitchers. When moving about (which they do more than dowitchers), Stilts walk deliberately with a down-pointing bill; dowitchers rarely show such a long, tapered silhouette and often move forward with their bills still underwater.

■ **Peeps** Peeps are the smallest members of the sandpiper genus *Calidris*. Three species are common in North America: Semipalmated, Western, and Least. Four additional species occur as rarities from the Old World (known as *stints*). The identification of the Old World species depends on a thorough knowledge of our common species and requires careful looks at feather details. Check your field guide or one of the specialty shorebird books for full details.

As with most sandpipers, there are three different plumages to

A single photograph can't capture the nuances of movement and posture, but notice **Stilt Sandpiper**'s longer legs, compared to the dowitcher: You can tell they're longer because there is much more space between the belly and the water. Both Stilts are moving around with their bills in typical, down-pointing position, keeping them just out of the water. The plump and colorful juvenile **Short-billed Dowitcher** on the left is obviously different, but basic-plumaged adult dowitchers can resemble Stilt Sandpipers. The Stilt Sandpiper in the center is a juvenile molting into first-winter plumage—notice the juvenal wing. The Stilt on the right is an adult molting into winter plumage—notice the barring on the belly. (*New York, September*)

juvenile Short-billed Dowitcher

juvenile Stilt Sandpiper molting into first-winter plumage

adult Stilt Sandpiper molting into winter plumage

This group shot features **Western** *(left)*, **Semi-palmated** *(center)*, and **Least Sandpiper** *(right)*. Even though all three species are close together, the Least is actively nearer the vegetation; feeding Westerns and Semipalmateds favor open mudflats or very shallow water. Notice the juvenile Western's paler head and rufous-edged inner scapulars; the juvenile Semipalmated has grayish, more uniform upperparts with a scaly look. Adult Least is much darker (when worn like this one) and browner than the others. *(New Jersey, August)*

learn: breeding adult, nonbreeding adult, and juvenal. For these three species, then, that makes a total of nine different plumages. Additionally, birds in molt look somewhere in between; worn birds (ones that have worn off the pale fringes to many feathers) also look different. It's no wonder that so many beginners shake their heads and move on. It can help to take a holistic approach and consider some preliminary ways of sorting them out, such as feeding behavior and shape, for example.

These three species have some general feeding tendencies that can be a good starting point, after which you can move on to structure and finally feather details. Least Sandpipers favor some sort of vegetation cover, and the cover can be right next to an open mudflat; but in large mixed flocks, Leasts tend to be at the edge nearest any vegetation or actually in the vegetation. Semipalmateds and Westerns tend to feed in more open situations—on wet mudflats or in very shallow water.

■ **Black-and-white Warbler** Some wood-warblers don't fit the mold. This species' behavior is almost as distinctive as its plumage. Black-and-white Warblers act like nuthatches, employing the same vertical creeping behavior while foraging. They rarely ascend to the leafy tops of trees or perch on the ground; they prefer the midlevel trunks and larger branches. The Black-and-white Warbler moves about in an animated mechanical fashion and often clings to the bark upside down as it peers into crevices; its elongated hind claws help it to hold on. Its fine-tipped, slightly curved bill is perfect for extracting insects hidden in bark crevices, a food source available earlier in spring than most flying insects and most likely a factor in this species' early arrival in spring.

■ **Courtship Displays** This behavior can be elaborate and ritualized in some birds, nonexistent in others. Some nonpasserine species have evolved exotic displays that birders make special trips to observe. The two prairie-chickens, Sharp-tailed Grouse, and the two sage-grouse are particularly popular. These species "dance" and strut at dawn on traditional display sites *(leks),* where the males compete with each other for mating privileges with the onlooking females. Other displays serve to solidify the pair bond prior to breeding, like the "rushing ceremony" and "weed ceremony" of Western and Clark's Grebes, shown on page 116. *(See also photograph of Double-crested Cormorants on page 69.)* These are examples of spectacular and very noticeable displays, but the vast majority of species exhibit displays that are brief and often not recognized by birders for what they are. The individual species accounts published in *The Birds of North America* and also available online are a reliable resource for additional information.

■ **The Unexpected** This is the what-the-heck-is-that-bird-doing category. If you get out in the field enough, you'll see the unexpected fairly often. Our perception of what is unusual is based on our experience, not necessarily on what is rare or uncommon. The fact that Northern Mockingbirds often sing at night qualifies as unusual behavior to the nonbirder. Even with birds we think we know well, rarely observed behavior could, in fact, be quite common. Bitterns are rarely observed swimming *(see page 117),* but related herons and egrets occasionally alight on deep water and swim briefly before taking off. Maybe bitterns swim more often than we think. Regardless of how truly rare or uncommon a behavior is, if it's surprising to you, it's worth noting in your journal.

This pose says it all: Among the wood-warblers, only **Black-and-white Warbler** forages in this fashion. This female was photographed in California, where it is one of the most numerous "eastern" warblers found there in fall. *(California, September)*

Voice

The vocalizations of birds stir strong feeling in most of us. You don't need to be a birder to be charmed by the simple melodies and varied call notes of the birds in your backyard. In wilder places, the bugling of a flock of Sandhill Cranes, the plaintive whistle of a migrating plover, or the dawn chorus on a spring morning make for vivid memories.

For all the poets' words and daydreaming that birdsong inspires, birds sing for practical reasons, none of which have to do with *Homo sapiens.* For the most part, male songbirds are the ones that sing; they do so to establish and defend territory and to attract and hold a mate. Calls are used for a variety of communications and,

Western Grebes (left) engage in complex and energetic courtship displays. This shows part of the "rushing ceremony": Raised wings are left folded; the triangular space below is filled by the spread scapulars; necks are arched forward; and lobed feet propel the bird through the water.

During display the male **Greater Sage-Grouse** (right) inflates his air sacs and jerks his head downward into his white ruff, producing a very loud, liquid popping sound that can be heard up to three miles away. The audible part of the strutting display typically starts with two wing swishes, followed by three low-frequency *coos* and then two air sac *pops* separated by a whistle, and tapers off with several short hoots. Amazing.
(left: Manitoba, May; right: California, April)

unlike most songs, they are given year-round and by both sexes. For birders there is great practical value in the study and recognition of birdsong. Its undeniable beauty only adds to the interest.

Starting Out Bird vocalization is often the first clue to a species' presence and a directional clue to its location. In many situations, birds' songs and calls are the primary clues telling you what is in the neighborhood and whether there is anything unusual to look for—if you can process the clues. If you're on a field trip, ask questions about what you're hearing. "Did you hear that?" and "What made that sound?" are *not* dumb questions. Don't hesitate to ask questions; you'll learn faster by asking for help.

In most birding locations, the first thing to do is to stop and listen. Take an auditory inventory of your surroundings. Maybe you know next to nothing about what species is making what sound. That's fine. You can still get an idea of where to search for it. If you hear something you don't recognize, locate it, identify it visually, and then watch it until it sings or calls again. That experience, repeated often enough, cements the sound with the image and the name.

Most people in the world hear and enjoy the songs of birds but give no thought to what species are making those songs and how the songs differ. Learning to identify bird songs takes effort. Everyone can learn to recognize some bird songs, and any effort you make will boost your birding skills as well as your enjoyment. Much like practicing an instrument, you need to train and exercise your ears, tuning in to the natural world. The key to learning is repetition. Hear, locate, and identify; listen as long as you can to anything new or unfamiliar. Make the visual and aural connection between the bird and its song.

Songs Nearly half of the world's species are popularly known as songbirds (technically, these are called oscine passerines) which in North America include all the passerines except the flycatchers. Songbirds comprise the species located in the back half of your field guide. They have developed more complex and varied songs than all other bird groups, and they produce the music that humans appreciate.

As noted, for most songbirds the male is the singer and his primary song serves as a territorial declaration, but for a few species—for example, Northern Cardinal and Baltimore Oriole—the female also sings. Some species change songs: The song used for establishing territory and attracting a mate becomes a slightly different song once the pair is established. Other species have slightly different songs depending on whether they are in the center of their territory or the edge. Some species, such as Northern Mockingbird and Brown Thrasher, have hundreds of variations on their primary song *(song repertoires)* but can be identified by the overall pattern of their songs. Brown Thrasher repeats phrases twice and moves on to a different phrase; Northern Mockingbird repeats phrases four or more times; and the related Gray Catbird gives a variety of phrases but without repetition. A few songbirds have very limited repertoires: For example, Cedar Waxwings produce high-pitched *zee* notes, which sound only slightly more impressive when given by a flock.

Other less well-known song types are part of many songbirds' repertoires. The *dawn song* is a variation heard by early risers and is often more complex than the primary daytime song. Sometimes birds give a very quiet song known as a *whisper song*. The whisper song can be a soft rendition of the primary song or something entirely different. It's most often heard on migration and at midday or during inclement weather.

Is that **American Bittern** *swimming*? It's not what you would expect to see, but many species that live around water can swim in a pinch. (Texas, March)

A male **Black-throated Green Warbler** holds forth. This species has two well-known songs. The first is for establishing a territory and attracting a mate: *zoo zee zoo zoo zee (trees, trees, whisp'ring trees).* The second song emerges after the pair is established: *zee zee zee zoo zee (see, see, see, Suzie).* (Michigan, May)

As with human speech, regional *dialects* occur in some birds. Non-migratory species that have little contact with other populations are the most prone to develop recognizable differences. The sedentary, coastal race of White-crowned Sparrow in California *(nuttalli)* sings different song dialects in adjacent valleys!

Some species are excellent *mimics,* incorporating other birds' songs and even nonbird sounds into their songs. Northern Mockingbird and Brown Thrasher are well-known examples, but European Starling, Lesser and Lawrence's Goldfinches, Yellow-breasted Chat, and others are also excellent mimics. Some species, like Blue Jays and starlings, mimic the calls of other species: Where Blue Jays occur, birders often get fooled by their imitations of Red-shouldered or Red-tailed Hawks.

Calls Calls are short, standardized vocalizations produced by all birds, and they are the primary vocalizations of most nonpasserines. Unlike most songbird songs, calls are innate—that is, they are not learned—and are given year-round and by both sexes, which make them very important and reliable identification aids. Among the variety of call types, the most commonly heard are *contact calls* (or just calls). These calls are used to maintain contact not only with a mate but also with flocks of the same or different species. For example, the mixed flocks of songbirds common in winter and during migration keep together using contact calls.

Even though calls are brief, they can be distinctive and important to identification. The calls of the secretive rails are often the only indication of their presence, the two dowitchers are most easily separated by calls (not plumage), and finding nocturnal species such as owls and goatsuckers depends on hearing their calls.

Although the contact calls of many songbirds (also called *chip notes*) sound similar to the untrained ear, most differ enough to be useful for identification. It takes time and experience to distinguish (and remember) the flat *chip* of a Chestnut-sided Warbler or the sharp *dit* of a Black-throated Blue Warbler. Chip notes can be very helpful even if you're a beginner: While you're hearing the chip notes of common birds, train your ear to listen for anything different and track it down. A different chip note will often be your first clue that a different species is nearby.

Many songbirds give *flight calls* during their nocturnal migrations or simply when they fly from perch to perch. These calls differ in tone and intensity from the standard contact calls described above. On favorable migration nights (windless nights are best for listening), experienced listeners can identify various species. Thrushes, warblers, and sparrows are often heard on good flight nights. Remote monitoring of nocturnal flight calls is used to monitor songbird migration in a more scientific way; computer analysis can identify the overflying birds to species and count numbers as well.

Other types of calls are occasionally heard. The most common are *alarm calls*, which resemble louder, more animated contact calls. Some species give a flocking alarm call that warns of a nearby predator: Bushtits give high-pitched, descending *didididididididi* notes; if you hear this distinctive sound, check the area for a raptor. Young birds give *begging calls*, even after they have left the nest. Young songbirds

The very similar eastern **Willow Flycatcher** *(left)* and **Alder Flycatcher** *(right)*, which belong to the genus *Empidonax*, are best distinguished by songs and calls. Alder sings a falling wheezy *weeb-ew* and gives a loud *pip* call; Willow sings a sneezy *fitz-bew* and gives a liquid *wit* call. Subtle visual clues include Alder's more distinct eye ring, brighter tertial edges and wing bars, and slightly smaller bill. *(left: Ohio, May; right: Ohio, July)*

The look-alike **Fish Crow** *(top)* and **American Crow** *(bottom)* are almost impossible to tell apart visually. Fortunately the familiar *caw, caw, caw* calls of the American Crow are very different from the nasal *uh uh* calls of the Fish Crow, but juvenile American Crows give begging calls that can be confusing. A few visual clues may help a little: The smaller Fish Crow has shorter legs, more pointed wings, and quicker wing beats; and the iridescence on the back is more uniform, lacking the scalloped or ringed pattern on American. *(top: Connecticut, February; bottom: Connecticut, December)*

often follow their parents around making unmusical whining calls. Juvenile terns accompany their parents for weeks after fledging and make some of the most distinctive begging calls *(see page 123)*.

Nonvocal Sounds Not all bird sounds are vocalizations. A number of species have evolved other ways of producing sound. The ritualized *drumming* of woodpeckers (as opposed to their hammering on trees to excavate food or nest holes) functions as a territorial song. The pattern, speed, and loudness of this territorial drumming vary among species and can be useful for identification.

The wings of some species produce distinctive sounds. During flight, the narrowed shape of the outer primaries produces the trilling sound made by some adult male hummingbirds. The dry wing buzz of Black-chinned Hummingbird is much different from the musical wing trill of Broad-tailed Hummingbird. The "booming" sound of displaying Common Nighthawks is caused by air displacement at the bottom of a dive. The whistling sound of Mourning Dove's wings on takeoff is different from the loud clapping sound made by Band-tailed Pigeons.

Many species of grouse have evolved complex courtship rituals, some of which include nonvocal sounds. The "drumming" of Ruffed Grouse and loud wing clapping of Spruce Grouse are examples. Clacking noises made by quickly snapping the bill shut are part of many albatross courtship rituals. The narrowed tail feathers of Wilson's Snipe, which are extended at right angles during its display flight, produce a hollow, owl-like whistle (known as *winnowing*).

Describing Songs and Calls Like any music, bird songs and calls can be thought of and discussed with reference to tempo, pitch, and loudness. Most people have a basic understanding of what the terms *fast tempo*, *high-pitched*, and *loud* mean, but without a reference point, the information is somewhat vague—faster, higher, and louder than what? Still, these generalizations are useful for describing the basics of a specific song. If possible, compare the song or call to a similar one with which you are familiar. As you listen to songs, pay close attention to the pattern of the notes and the length of the pauses between phrases. These give a song its tempo and rhythm.

The description of the *quality* of a song is often very useful. You might say that a song "is composed of liquid, flutelike notes that spiral downward in pitch." A description like this sounds somewhat vague, but the liquid, flutelike notes are a signature sound of the brown thrushes in the genus *Catharus*, although only one species in that group, Veery, has such a distinctive downward spiral to its song. Getting to this point assumes some experience with brown thrushes, but as these species are not uncommon, beginners soon recognize the unique quality of their songs.

Another approach is to use visual annotations for taking written notes. A well-known system uses long and short dashes to denote time; placement of the dashes at higher or lower positions denote pitch. For example, the long, trilling song (fast notes all on one pitch) of Chipping Sparrow could be represented as follows: - - - - - - - - - - - - and Black-throated Green Warbler's buzzy *zee zee zee zoo zee* song might be — — — _ ⎯ , with the fourth note lower and fifth note higher in pitch. It's not a bad way to get something down on paper, and it forces you to analyze what you're hearing. In the latter example, adding the phonetic syllables *(zoo* and *zee)* and a description of the song's quality (buzzy) fills out the description.

Scientific representations of bird songs, called *sonograms,* are quantitative graphs of sounds. The horizontal axis represents time and the vertical axis represents pitch (or frequency). Straight horizontal lines represent clear tones, and vertical lines represent buzzy tones with a wide frequency range. Sonograms, which are essential tools for song researchers, are becoming more widely appreciated by birders, although it takes some effort to get familiar with them. The Golden Field Guides' *Birds of North America,* first published in 1966, is the only popular field guide that includes sonograms. You can also find free software on the Internet that allows you to make your own sonograms.

■ **Memory Aids** The repeated experience of seeing and hearing birds in the field will have the greatest effect on your ability to recall songs and calls, but some memory aids are also helpful.

There's a tradition in birding of inventing catchy phrases as *mnemonics* or memory aids—phrases that capture the tempo and sound of a species' song. Most birders learn some of them, and the field guides usually mention the time-tested favorites: For White-throated Sparrow we have *Old Sam Peabody, Peabody, Peabody* or *Pure sweet Canada, Canada, Canada,* and for the two different songs of Black-throated Green Warbler, we have *see, see, see, Suzie* and *trees, trees, whisp'ring trees.* The best mnemonics combine the sound and tempo of the song with something about the bird: Most White-throated Sparrows nest in Canada; Black-throated Green Warblers often nest in coniferous forest (where presumably the trees are whispering). Other mnemonics like Olive-sided Flycatcher's *Quick-three-beers!* and Eastern Towhee's *Drink-your-tea!* are just plain memorable. Some birders even invent their own mnemonics. To one person's ears, Carolina Wrens seem to be saying *do-research, do-research,* to another *do-it-again, do-it-again;* and at least one field guide writer heard *teakettle, teakettle.* Whatever works for you is what to use. Other birders prefer to stick with written-out phonetic sounds, like *zee zee zee zoo zee* instead of *see, see, see, Suzie.*

Another memory aid is built right in to some species' names: These names were created from the sounds the species make *(onomatopoeia,* as in the word *sizzle).* Some examples are bobwhite, chachalaca, Killdeer, curlew, Willet, Whip-poor-will, Chuck-will's-widow, poorwill, phoebe, pewee, kiskadee, chickadee, and Dickcissel. Occasionally a name will lead you down the wrong path: Screech-owls, for instance, don't screech; in fact, all three North American species have rather mellifluous voices.

Some memorable descriptions of song quality have entered the birding lexicon. The origin of many of these is shrouded in time, but Roger

Tory Peterson is credited with a few of the most unforgettable ones: Scarlet Tanager sings like "a robin with a sore throat," and Rose-breasted Grosbeak sings like "a robin that has taken voice lessons." These catchy descriptions of the quality of a song will help you recall it easily.

■ **Recordings** Commercial audio cassettes and CDs are a great way to supplement your field experience. Limit your listening to a few songs, or they'll blend together and be more confusing than helpful. A good approach is to review songs you've recently heard in the field. If you're listening to a CD, set it on repeat so the track plays over at least a few times. Many birders enjoy refreshing their memories in early spring by listening to tracks of warblers singing to remind themselves what Magnolia Warbler sounds like, for example. You can test your knowledge by constructing the personalized quizzes found on some websites.

Inexpensive minicassette recorders work quite well for making your own recordings. Identify the species and location on each recording; otherwise, your recorded snippets will be very difficult to keep straight. Digital recorders and MP3 players have excellent sound quality and are small and portable. Some of these digital products can be purchased with preloaded recordings, or you can transfer recordings from CDs.

Many birders use their own or prerecorded tapes to attract birds. The birds, usually males, assume they are hearing a territorial rival and come to investigate. This is an effective technique but should always be used sparingly so as not to disturb nesting species.

The begging calls of a juvenile **Caspian Tern** are high-pitched whistled notes, completely different from the harsh throaty scream of an adult. Both calls are very loud, frequently given, and carry long distances. Immature Caspian Terns continue to give juvenile-like calls through the following spring migration, when they are close to a year old. Juvenile Caspians look and sound so different from adults that unsuspecting birders often assume they must be a different species. *(Ohio, September)*

CHAPTER 6

VARIATION IN BIRDS

B irding would be much easier if every individual of a species looked the same. Learn one figure for each species, and you'd be done. Your field guide would be a slender volume with only one image per species. However, a lot of the fun and challenge of bird identification and insights into bird biology would be lost.

For a few of our species, there are only negligible differences between sexes and among age groups and geographical populations. Consider the widespread and well-known Blue Jay. The sexes appear identical in the field, and the fifth edition of the *National Geographic Field Guide to the Birds of North America* shows only one illustration. *Aging* Blue Jays—differentiating between Blue Jays of different ages—can be done with only very close views, usually in the hand by bird banders. Although there are three named subspecies of Blue Jay, the variation is very slight and clinal, and they are not field identifiable. This obliging bird looks almost the same wherever and whenever you see it.

Blue Jay, however, is the exception; most species of birds are variable in one or even multiple ways. In this chap-

Blue Jay *(above)* shows very little variation— geographic, sexual, or age related. Essentially, all Blue Jays look alike. At the other end of the spectrum, **Williamson's Sapsucker** *(opposite)* exhibits such striking sexual differences that the male *(left)* and female *(right)* were originally thought to be separate species. *(above: Michigan, January; both opposite: Colorado, June)*

ter we will explore that variation and how understanding variation is important to your maturation as a birder. *Sexual variation* (or sexual dimorphism) is perhaps the easiest to understand: Males and females look different. The process of molt, the regular replacement of feathers, often produces other types of variation. Birds that are actively molting can appear different, and wear and fading of old feathers can produce changes in appearance as well. For some species, *seasonal variation*—looking different at different times of the year—is very marked. We describe these species as having an alternate (breeding) and a basic (nonbreeding) plumage. *Age variation*—looking different at different stages of life—is produced when birds molt in a succession of differ- ent plumages before reaching adulthood; plumage differences by age class are perhaps most well known in the large gulls.

Geographic variation occurs in most species in North America, especially with our passerines and particularly in the West. This variation can be expressed in many ways—size, color, voice, to name a few—with variation among populations being much greater than that within populations. These different populations of the same species are often recognized by the naming of *subspecies,* or *races.* Named subspecies, especially geographically adjacent ones, can be difficult or impossible to tell apart in the field; they are often placed in *subspecies groups.*

Other types of variation are less common but potentially confusing. Color morphs, individual variation, and hybrids between two species will be discussed at the end of the chapter. Even aberrant plumages, such as albinos or leucistic (abnormally pale) birds, are occasionally seen. All of this variation is part of what makes the world of birds so interesting. Once you begin to familiarize yourself with the range and frequency of this variation, you'll enjoy observing it in the birds you see.

Sexual Variation

Williamson's Sapsucker of western North America exemplifies a species with striking sexual variation. For several decades in the latter half of the 19th century, most eminent ornithologists believed that the males and females were two different species! The female was described to science by the famous John Cassin (after whom species of auklet, kingbird, vireo, sparrow, and finch are named) in 1852, and the male was discovered and described as a different species five years later. This flawed treatment was adopted by other leading ornithologists (including Spencer Baird, William Brewer, and Robert Ridgway), and it remained so until 1875, when Henry Wetherbee Henshaw solved the riddle. This confounding of the ornithological community was the exception, of course, but many of our species show distinct sexual variation. In certain groups, including waterfowl and a number of passerine families, such as wood-warblers, tanagers, and blackbirds, males and females differ strikingly. For others the plumage differences are less striking and more careful observation is required.

For species like our Blue Jay and many others—including gulls, terns, owls, most flycatchers, crows and jays, chickadees, and titmice—the sexes appear basically identical in the field. Keep in mind that there are usually slight to occasionally obvious *size differences* between the sexes. Males are usually, but not always, larger; females are distinctly larger in raptors and some shorebirds, especially phalaropes. For many species there is only a slight average difference, and *sexing* (determining the sex of) a bird might only be possible in hand by a taking a series of measurements. For some species (hawks, owls, gulls), size difference becomes most apparent when viewing a mated pair. During copulation, an affair that lasts only a few seconds in most species, males are always on top.

For species with distinct songs, like our perching birds or passerines, the male is usually the one that sings. (A well-known exception is Northern Cardinal.) If a bird you thought was a female suddenly bursts into song, reconsider your identification: It is probably a duller, immature male. Singing males are more conspicuous and easily seen during the breeding season than are females; they are also generally the first arrivals, both at migratory stopovers during spring migration and on the breeding grounds. Keep in mind that for some species where normally only the male sings, there are recorded instances of female song. Also, in known cases where the female sings, the song of the female can sharply differ from the male (e.g., Wrentit). In some instances the males and females tend to winter in different habitats or have slightly different winter ranges.

In species that exhibit sexual variation, what is discernible in the field varies among species. As seen on page 125, Williamson's Sapsucker shows dramatic differences, but in other species the differences can be *very* subtle. Additional examples of species with striking sexual variation include our North American tanagers (genus *Piranga*) and species like Rose-breasted Grosbeak and Indigo Bunting when they are in alternate (or breeding) plumage. Most of our wood-warblers show distinct

The three phalarope species exhibit sexual role reversal: The females are more colorful in breeding plumage, and the more somber-plumaged males incubate the eggs and care for the young. Note the more colorful female under the male on this pair of copulating **Wilson's Phalaropes**; many males are duller than this individual. (California, May)

These **Rose-breasted Grosbeaks** exhibit striking sexual differences. The male *(left)* is a full adult on the basis of wing and tail pattern. *(both: Texas, April)*

sexual variation, especially within the largest genus, *Dendroica*. The most highly differentiated wood-warbler is Black-throated Blue Warbler, where the males and females look like completely different species.

Sexing individual birds can be complicated by seasonal and age differences. What may be a striking sexual difference in spring and early summer can be a subtle to very subtle difference by fall (e.g., Yellow-rumped Warbler). Certain species, like Kentucky Warbler and Zone-tailed and Rough-legged Hawks, show only subtle sexual differences. Even more subtle sexual differences are found with species such as American Avocet, for which the best method of sexing is the shape of the bill—longer and straighter on the male. Even within the wood-warblers, where most species can be sexed in the field, males and females in some species appear identical, as is the case with Worm-eating and Swainson's Warblers.

In general, males are brighter in color. Some of the best-known exceptions are the three species of phalaropes (shorebirds), in which females are both larger and more colorful. Interestingly, phalaropes show strong role reversal of the sexes: Males take on the duties of incubating the eggs and accompanying the newly fledged young. Considering their different roles while breeding, the male's duller and more cryptic plumage is genetically well-adapted for avoiding predators.

Molts and Plumages

Molt is the process by which feathers are shed and replaced. In some species, molting also involves the shedding or the growth of horny sheaths, as on the bills of some of the alcids, such as puffins. This replacement is crucial for survival, because feathers are exposed to many environmental agents including sun, physical damage (such as that done by vegetation), and parasites and bacteria—all of which degrade the overall condition of the feather. Feather degradation can also impede flight, and in waterbirds the waterproofing and insulative properties of the feathers can be lost. Feathers are a key evolutionary innovation, and their maintenance and regular replacement is vital to birds. Most field guide illustrations and many photographs of birds show them in pristine plumage because publications tend to avoid illustrating molting birds that look incomplete or messy. But birders who are out year-round will sooner or later see birds that are heavily worn or actively molting.

Studying molt can be an obsession for some, and many important research papers and reviews on the subject have been written recently. For most birders, however, the brain fogs over at the mention of molt. Nevertheless, birders cannot completely avoid the subject, for the study of molt can sometimes be used in field identification, as we will see in the examples cited below.

The Process At the base of the skin is a feather *follicle* at which the feather is attached. During a molt, the *dermal papilla* at the base of the follicle is stimulated, producing the growth of a new feather, which emerges and pushes out the old feather. Initially the new feather is sheathed, but rapidly the hard plasticlike outer substance breaks away, revealing the new fresh feather. Sometimes a feather or even groups of feathers may be lost suddenly through an accident, like a near-death encounter with a predator. These feathers are usually replaced quite rapidly.

Normal molt occurs twice a year in many species: a complete molt at the end of the breeding season or during fall, and a partial molt prior to the breeding season. As the name implies, the complete molt involves the replacement of all of the feathers, including the flight feathers, which are so important for survival. The process is gradual and orderly. Not all feathers are shed at once, for obvious reasons. Most species retain the power of flight when molting (though some waterfowl, for example, are rendered temporarily flightless during their complete postbreeding molt). Within the flight feathers, usually the innermost primary and outermost secondary are dropped first, then sequentially the next primary out and the next secondary in, until finally the outermost primary and the innermost (next to the body) secondary are replaced. The complete molt usually takes more than a month among our passerines and still longer for others. For gulls it takes from a few months to nearly half a year. Some studies indicate that within the passerines, at least, migratory species molt more quickly than sedentary ones, even within different populations (or subspecies) of the same species (for instance, within Old World White Wagtail). In addition to the complete molt, many species have a second, less-complete molt in spring involving body feathers but not flight feathers. This is often when the plumage becomes more colorful, enhancing key reproductive behaviors such as territory establishment and defense as well as mate attraction. Two North American species (Franklin's Gull and Bobolink) have two complete molts a year. In many very large birds (pelicans and albatrosses, for instance) the complete molt takes even longer, and two or more simultaneous "waves" of molting flight feathers can be detected.

A freshly molted **European Starling** in fall and winter *(top)* is heavily spotted with buff. The buff feather tips become brittle and wear off by spring *(bottom)*, revealing the glossy colors previously hidden. *(top: Connecticut, December; bottom: Texas, June)*

Molt Terminology We have two widely used systems to describe the molt patterns and plumages for birds.

■ **Life-Year System** This system describes a bird's appearance during the course of the year and is not necessarily based on molt: for example, *juvenal, first-winter, first-summer, first-year, second-winter, second-summer, second-year*, and so on, until *adult* plumage is finally attained. The terms used depend on the species or family in question. For instance, a five-month-old White-crowned Sparrow in November has replaced its juvenal body feathering and will be called a first-winter bird, whereas a Bald Eagle won't show its first postjuvenal plumage until well into its second calendar year.

■ **Humphrey-Parkes System** In 1959 a more comprehensive approach to molt terminology was proposed by Philip S. Humphrey and Kenneth C. Parkes, one that could be used throughout the world, rather than being biased to more temperate northern latitudes. In this system, every named plumage has resulted only from molt *(see table opposite)*.

All species have a *basic* plumage, and for those species that have only one plumage after juvenal plumage, their molts produce only one basic plumage after another. As we noted, many of our species exhibit a *partial* molt (*prealternate* molt) and acquire an *alternate* (or *breeding*) plumage. This terminology works fine for identifying adults, but additional terms are used for the younger birds. The *juvenal* plumage is the first true set of feathers after the downy or nonpennaceous feathers (plumages only seen in or near the nest). This plumage is retained for a variable period of time: only a few weeks for our wood-warblers, a year or so for many of our raptors. (An individual in juvenal plumage is referred to as a *juvenile*.) The *prebasic 1* molt produces the *basic 1* plumage, and the *prealternate 1* molt produces the *alternate 1* plumage, until the adult plumage is reached. *Definitive* plumage is one that no longer changes with age. It is associated with adult plumage, hence the terms *definitive basic* and *definitive alternate*. Some species have additional intermediate plumages, known as *supplemental plumages*, and the molt that produces them is known as *presupplemental*. Among the species with supplemental plumages prior to the winter season are most of our *Passerina* buntings (Lazuli, Indigo, Varied, and Painted).

It is important to remember that under the Humphrey-Parkes system, plumages are produced only by molts, not by wear.

Molt	Plumage
Prebasic	Basic
Prealternate	Alternate

Thus the male Snow Bunting in his black-and-white plumage in spring and summer is still in basic plumage, as this plumage was acquired completely from wear. The same applies to that smart-looking male House Sparrow in spring. On European Starlings *(left)*, the buff spots on the feathers acquired from the prebasic molt in late summer become brittle and break off in late winter or early spring, producing the purplish color of breeding birds. Recent studies have concluded that Lawrence's Goldfinch acquires its colorful breeding plumage (again, technically not alternate plumage) entirely through wear. Some species (e.g., longspurs) acquire a breeding-type plumage partly through molt and partly by wear.

This overview is only a very general one; for the most part, complexities are not addressed here. For instance, in the larger gull species, especially those that take nearly four years to acquire definitive adult plumage, individuals seem to be in a nearly constant state of molt; for the younger birds, depending on when a feather is dropped during the year, the new feather may be a different pattern and color, even with birds of approximately the same age. For simplification, some have proposed using a terminology based on cycles. Thus a gull in its first year of life would be in its first cycle, during its second year it would be in a second cycle, and so on, until it becomes an adult.

The following two age terms are widely used by birders, though they

are not formally a part of the Life-Year or Humphrey-Parkes systems.
■ **Immature** This term applies to any bird that is less than a full adult.
It is often, but not always, used for birds older than juveniles.
■ **Subadult** This term refers to a bird at some intermediate age between
a juvenile and a full adult.

Timing of Molt The molting process depletes the bird's energy reserves.
Certainly most species do not molt while actually migrating, a high-
energy activity. The prebasic molt usually takes place either on the breeding grounds, prior to fall migration, or on or near the winter grounds.
Some species start the prebasic molt on the breeding grounds and then
suspend it, completing it once the winter grounds are reached.

Different molting strategies reflect different conditions. Consider four
eastern passerines: Scarlet Tanager, Baltimore Oriole, Indigo Bunting,
and Rose-breasted Grosbeak. These species remain on their breeding
grounds to perform their prebasic molt and then migrate south in basic
plumage. Contrastingly, most adults of their western counterparts (Western Tanager, Bullock's Oriole, Lazuli Bunting, and Black-headed Grosbeak, respectively) migrate south first and then go through their prebasic
molt, usually on the winter grounds or (for many Lazuli Buntings, among
others) in an intermediate region. For the eastern species cited above,
their fall migration is correspondingly later than that of their western
counterparts, at least for the adults. These contrasting strategies have
evolved in response to differing environments. Consider that the East
is usually rich with food resources after an abundance of summer rains;
over much of the West, however, it has been extremely dry since the
winter rains and spring snowmelt. Molting when food resources are in
short supply is not a good strategy. Lazuli Buntings that undergo their
prebasic molt in the Southwest in August and September take advantage of abundant food crops that appear after the summer rains have
started in early July.

Most adult shorebirds migrate south in late summer in their alternate plumage (now more worn) and don't molt until their winter
grounds are reached or approached. A few, like Purple Sandpiper,
molt into their basic plumage on the breeding grounds. This is also
true of those Dunlin subspecies that nest in northern Siberia, the
Russian Far East, and North America, excluding Greenland. They
are, therefore, among our later shorebirds to arrive in fall in southern Canada and the lower 48. But those Dunlins nesting on the Arctic coast of western Russia, northern Europe, and Greenland migrate
south earlier in their alternate plumage and then molt. Other species
(e.g., phalaropes) migrate partway south, molting at a staging area

with highly productive food resources. For Wilson's Phalarope, Mono Lake in eastern California is one such well-known staging area, and the edge of the lake in late summer is littered with the feathers of both adults and juveniles.

The prealternate molt usually takes place on or near the winter grounds; thus in spring, when we first see these species, they are usually in alternate plumage. This is a generalization, of course, and some early migrants might still be in a basic-like plumage. In other cases this is true for the species as a whole. When American Golden-Plovers arrive in the western Gulf Coast region in March, all are still in basic plumage. Their migration continues northward, and birds seen at the end of March and early April in the upper Midwest are still in basic plumage. Yet those migrants arriving in the upper Midwest from late April to mid-May are in their full, stunning alternate plumage.

As most birders well know, looking at summer and early fall ducks is a study of shades of brown, the males being nearly as dull as the females. Male ducks actually molt into their alternate plumage during fall and do their courtship on their winter grounds or during the spring months. Their prebasic molt occurs during summer, when they briefly hold a brown *eclipse* plumage. This is true only for the males; the timing of the molt in female ducks differs.

Bander Terminology In North America hundreds of licensed bird-banders (known as *ringers* in the United Kingdom) catch birds in mist

Adult **Long-billed Dowitchers** on their southbound migration typically linger at large wetland sites, where they molt into their duller basic plumage. This adult is part way through its prebasic molt. Short-billed Dowitchers (*not pictured*), on the other hand, molt on or near the winter grounds. Thus, a heavily molting dowitcher seen inland in fall is almost certainly a Long-billed. (*California, August*)

worn breeding (alternate) feathers—barred pattern

fresh nonbreeding (basic) feathers—gray with central streak

nets or traps, record certain data, attach aluminum leg bands, and release the birds unharmed. Banders have long used age-coding terminology, which is now widely used by birders as well. This terminology is based on the calendar year:

■ **HY (Hatching Year)** This includes birds in juvenal plumage (separated by many with a separate code, *juv*) and is applied to birds up through December 31.

■ **AHY (After Hatching Year)** This bird is *at least* in its second calendar year.

■ **SY (Second Year)** This bird is in its second calendar year from January 1 through December 31; that is, its age is roughly six months to a year and a half.

■ **ASY (After Second Year)** This is an adult that is at least in its third calendar year.

■ **TY (Third Year)** This bird is in its third calendar year.

■ **ATY (After Third Year)** This bird is at least in its fourth calendar year.

■ **U (Unknown)**

The key thing to remember is that January 1 is the date for advancing to the next code. A small stumbling block for beginners is remembering that the designation *second year,* or *SY,* may apply to a bird that is only six months old or younger, not to a bird that is two years old.

Using Molt as an Identification Aid

Birds in alternate or breeding plumage often appear their brightest, but parts of their plumage (especially the flight feathers) may appear rather worn because these feathers—especially primaries and secondaries, but often some of the rectrices and tertials as well—were retained from a

These two similar species of *Empidonax* have different molt schedules. **Hammond's Flycatcher** *(left)* molts on the breeding grounds, and fall birds are correspondingly bright and in fresh plumage. In addition, the fall migration of Hammond's is somewhat later. **Dusky Flycatcher** *(right)* molts after reaching the winter grounds; thus birds seen in fall are dull and worn. Note Dusky's longer bill and tail and Hammond's rather long primary projection. *(left: California, September; right: California, August)*

Hammond's Flycatcher in fresh plumage

Dusky Flycatcher in worn plumage

immature
Common Terns

juvenile
Black Tern

adult Common Tern
in alternate plumage

adult Forster's Tern
in basic plumage

molt that occurred six to eight months earlier. Most species appear freshest and least worn after the prebasic molt in late summer or early fall, even though the overall colors may be more subdued. This may be because the actual feather is duller (as with many of our *Dendroica* wood-warblers) or because the brightly colored feather is veiled by a duller tip or fringe (longspurs). Sometimes, though, the overall effect produces a brighter bird by fall. Compare a freshly molted Grasshopper Sparrow in October with a worn July bird. The fresh fall bird could be easily confused with its buffy *congener* (another species of the same genus), Le Conte's Sparrow, if the head pattern is not examined critically. A kingbird with a forked tail is not necessarily a Tropical or a Couch's Kingbird, species known for their notched tails; it could be a molting Cassin's Kingbird that has dropped its central pair of tail feathers.

Knowing the timing of the molt often reveals important clues. For instance, the adult Long-billed Dowitcher pictured on page 133 is undergoing a rather extensive prebasic molt from definitive alternate to definitive basic plumage. We now know that Long-billed Dowitchers, unlike most shorebirds, often stage at large wetland sites to undergo their prebasic molt and then continue south. Adult Short-billed Dowitchers fit the more traditional model and molt once they reach the winter grounds, which are virtually *always* on the coast. Thus an inland adult dowitcher in extensive molt is almost certainly a Long-billed.

The Southwest has more species of hummingbirds than any other region in North America, north of Mexico. While adult males can be fairly easily separated in most cases, the species identification of females and immatures is much more difficult. A female Anna's (especially the immature) can look quite similar to a female Black-chinned or Costa's. During summer, one quickly assessed clue is whether the primaries are being molted. If primary molt is visible, Black-chinned and its

This mixed group of terns in early fall reveals important clues about identifying *Sterna* terns. The adult **Forster's Tern** (center)—note the juvenile **Black Tern** just *behind* it—has already molted into basic plumage, while the adult **Common Terns** (right) are still in breeding plumage with their solid black caps and gray-shaded underparts. They won't molt into basic plumage until well south of North America. The two terns on the left are immature Commons and can be easily separated from Forster's Tern by the presence of a dark carpal bar formed by the lesser and marginal wing coverts. (New Jersey, August)

congener Ruby-throated (casual in the Southwest) are eliminated, as those species don't start their primary molt until they reach the winter grounds south of the United States.

Carefully assessing molt also helps identify the notoriously difficult genus of flycatchers, *Empidonax*. Of the five eastern species, only Acadian Flycatcher remains on the breeding grounds to molt; the others migrate south, molting on the winter grounds. In the West, only Hammond's Flycatcher and the very restricted (in the U.S.) Buff-breasted Flycatcher remain on the breeding grounds to molt. This means that if one sees a heavily worn *Empidonax* in fall, Acadian and Hammond's are excluded. Acadian and especially Hammond's are also relatively later fall migrants than their North American congeners.

The genus *Sterna* comprises a number of medium-size terns, four of which occur in North America. All of these present their own identification problems; however, among the adults, Roseate and Forster's Terns lose (molt) their black caps by late summer while on the breeding grounds, whereas adult Common and Arctic Terns retain their black caps until after they have departed North American waters. Thus, if one sees a black-capped *Sterna* tern in September and October, it is either Common or Arctic Tern. Our two relatively widespread nighthawks, Common and Lesser, also show different molt strategies. Lesser Nighthawk molts on the breeding grounds, and by midsummer there are obvious gaps in the flight feathers, indicating dropped feathers. Common Nighthawk doesn't commence molt until reaching the winter grounds in South America. Thus, a summer nighthawk that is missing flight feathers (unless they've been lost accidentally, which almost never happens in a symmetrical fashion) is automatically a Lesser Nighthawk.

Pelagic observers off North Carolina have noted on late spring and early summer trips that adult Band-rumped Storm-Petrels are in flight-feather molt. This excludes the similar Leach's Storm-Petrel, which is not molting in that season. Of course a nonmolting bird could either be Leach's or a young (recently fledged) Band-rumped.

There are many other instances where knowledge of molt can reveal important clues about a species' identity. Although beginners may be tempted to avoid the entire subject, a more positive approach to learning about molt can help you immensely with field identification.

old outer primaries

missing primaries

new inner primaries

This **Lesser Nighthawk** is midway through its prebasic molt and can be easily separated from Common Nighthawk by this feature alone. Commons don't molt until after reaching their wintering grounds in South America in fall. The buffy bars visible on the inner primaries and the outermost 10th primary, which is slightly shorter than the 9th primary, are additional distinguishing features. The white bar is also closer to the tip of the wing. (California, July)

Age Variation Browse any field guide and you will learn that a good deal of the variation found within a species is a result of age. A quick scan of a sizeable flock of Double-crested Cormorants at any time of year is likely to reveal both black individuals (adults) and brownish or even whitish ones (immatures). A fall or winter flock of White-crowned Sparrows, a familiar sight at many western feeders, is likely to be composed of birds with black-and-white head stripes (adults) and tan-and-brown head stripes (immatures).

At first glance, age variation can seem bewildering. It's best to start by learning the juvenal plumage—the first true plumage—for each species, how long it is held, and what type of plumage is molted into next (definitive adult, or some intermediate stage). Passerines hold their juvenal plumage for only a short time. For instance, wood-warblers hold it for only a couple of weeks to a little over a month. A young American Redstart starts molting out of its juvenal plumage while still in the nest! With few exceptions (e.g., the odd, early migrant Yellow-rumped Warbler), juvenal plumage is only seen around the nesting grounds. Over much of North America, the spotted juvenal plumage of American Robin is a familiar sight, but the identification is pretty straightforward, especially if the parents are still feeding it!

In some species, juvenal plumage causes confusion. Almost every year well-out-of-range Sprague's Pipits are reported in late June, at a time when the superficially similar juvenile Horned Larks have just fledged. The streaked juvenile Brown-headed Cowbird (which will migrate in

Red-headed Wood-peckers exhibit striking age differences. The juvenile *(right)* maintains its brown plumage well into winter and then molts into a more adultlike plumage by spring, though many juveniles still have barred secondaries. *(both: Texas, July)*

this plumage) looks quite different from either adult parent and is routinely misidentified. The juvenile Painted Redstart lacks the red belly and at first glance looks odd, but it acts just like its parents and shows their distinctive wing and tail patterns. In most of our native sparrows, the juveniles are only seen on the breeding grounds, but there are exceptions—notably Black-throated Sparrows, Chipping Sparrows, and some young Le Conte's Sparrows. Juveniles of these species do migrate south and are noted well into October. Young *Myiarchus* flycatchers can migrate south in juvenal or partial juvenal plumage; Ash-throated Flycatcher does so routinely. The tail pattern, an important identification feature in this genus, is nearly useless for the identification of juveniles. Certain young woodpeckers, like the familiar Red-headed Woodpeckers of the East, migrate south in full or nearly full juvenal plumage *(see page 137)*. They will molt into an adultlike plumage during winter but often retain the juvenal wing feathers (dark bars across the white patches). Interestingly, young Yellow-bellied Sapsuckers also follow this pattern of retaining juvenal feathering into winter, yet similarly plumaged and closely related Red-naped Sapsuckers are largely adultlike when fall migrants are first noted in late September.

This young male **Scarlet Tanager** is hardly scarlet. As indicated by its scientific name—*Piranga olivacea*—its overall coloration is olive, a feature it shares with females and basic-plumaged males.
This bird has retained a few juvenal greater coverts that stand out as paler with whitish tips. A basic-plumaged adult male would be similar to this individual but would have solidly black wings.
(Connecticut, October)

dull brown
juvenal
secondaries
and primaries

For most passerines, the prebasic molt involves a replacement of the body feathers but not the flight feathers. For instance, all male Scarlet Tanagers in fall are largely greenish, but the adult males have entirely blackish wings, whereas immature males (first basic) have brownish primaries and secondaries that contrast with the black wing coverts. These juvenal flight feathers are retained for over a year, so in spring, despite the bird's scarlet plumage, the primaries and secondaries are still brown and now somewhat worn and faded. For most passerines, definitive adult plumage is acquired after their second prebasic molt, by their second fall.

In passerine species that show strong sexual dimorphism as adults, the sexes of the immature can often, but not always, be determined in the field as well. An exception involves two of our finches, Purple and Cassin's. For over a year, the younger males look just like females. They will hold territory and sing and actually breed in this plumage. Of course, the singing birds can be sexed as males, because females don't sing.

Waterbirds and hawks take longer to acquire adult plumage than passerines do. Hawks at nearly a year of age are still largely in juvenal plumage, having just started their prebasic molt. Juveniles of our familiar Common Loon can be told from basic-plumaged adults by both their duskier faces and the presence of pale tips (forming bars) on the scapulars. Their eyes are more brownish, rather than ruby red. By spring they will molt some body feathers, but the plumage is, overall, still like that of a basic-plumaged bird. Not until the loons reach nearly three years of age do they acquire the full alternate (breeding)

These two male **Scarlet Tanagers** are both in stunning alternate (breeding) plumage, but each represents a different age class. The first-spring male *(left)* retains its duller brown juvenal primaries and secondaries as well as the outermost tertial and outer tail feathers. On the full adult *(right)*, these feathers are solidly black. Migration timing also differs: The adult males usually arrive on the breeding grounds before the females and first-spring males. *(left: Texas, May; right: Ohio, April)*

plumage. With Bald Eagles, the process takes nearly five years; immatures and subadults present a dizzying array of plumages. As discussed in chapter 3, in most shorebirds the young migrate south in full, or nearly full, juvenal plumage later than the adults do *(see page 41)*. In some species of our larger gulls, the full juvenal plumage can be held well into winter; in others, molting starts by early fall. We are audience to a bewildering array of plumages for nearly four years for these larger gull species.

Awareness of the different plumages that result from age helps birders make an accurate identification. And aging is critical to identification in some families (including gulls and most shorebirds). Beyond just helping us make the correct identification, accurate aging has taught us much about the distributional and seasonal patterns of some species. In the Pacific states, for instance, small numbers of American Golden-Plovers appear in fall. Every well-documented record thus far is of a juvenile. Juvenile Long-billed Dowitchers are almost unknown in the lower 48 before September, yet juvenile Short-billed Dowitchers routinely appear in this region by early August.

Seasonal Variation Most species show some change in appearance through the seasons, whether obvious or subtle. As noted in our discussion of plumages and molts, birds are in fresher plumage in the fall after their prebasic molt. The prealternate molt in many species results in the acquisition of brighter colors that function in mate attraction, courtship, and territorial behavior. The widespread American Goldfinch is an example of a species that shows a distinct alternate plumage. With

Bald Eagles take about five years to reach full adult plumage *(right)*. Subadult birds *(left and top)* exhibit a variety of plumages, with a great deal of individual variation as they mature. *(Alaska, February)*

this change, which takes place largely during March and early April, the somber tones of gray and brown are replaced with yellow—brilliant yellow in the male. In some species only the male has a colorful alternate plumage; in others, such as American Goldfinch and many waterbirds, including Common Loon, both sexes acquire a distinct alternate plumage.

As explained earlier, this prealternate molt is a partial one and usually does not involve flight feathers. A quick review of any field guide will indicate which species show distinct seasonal changes. Check out the elegant alternate plumage of Common Loon and Red-necked Grebe and the colorful alternate plumage of the male Scarlet Tanager (see page 139), Bobolink, and Indigo Bunting. In other species, like Blue-gray Gnatcatcher, the changes are more subtle: The crown becomes more deeply blue, and a black line extends from above the eye across the forehead. Observers in the West, who aren't used to seeing black on the crown of this species, might be tempted to call it Black-tailed Gnatcatcher. Keep in mind that in many species, the male in its first alternate plumage acquires an intermediate plumage, one between that of the female and that of the older alternate adult male. An especially striking example of seasonal change is that shown by colorful spring wood-warblers, although in actuality the majority of our wood-warbler species don't show distinct seasonal change. However, some of our most numerous species within the genus *Dendroica* (including the abundant Yellow-rumped Warbler) do show significant changes between spring and fall.

These eastern **Blue-gray Gnatcatchers** can be easily sexed during the breeding season when the male (bottom) shows a distinct black line that extends from the forehead to beyond the eye. Birds seen in this plumage in the Southwest have often been confused with Black-tailed Gnatcatcher, which has a more solid black cap and different tail pattern. In winter (top) males and females are much more alike. (top: Texas, November; bottom: Ohio, April)

In addition to the acquisition of a colorful plumage, the seasonal change to breeding condition often involves changes in intense coloration of the bare parts (as in the facial skin of herons or in the orbital rings, feet, and bill of gulls), changes in intense coloration of horny bill sheaths (as in herons and gulls), or the development of either a horny sheath (some alcids) or a vertical epidermal plate (the "horn") along the top of the bill of American White Pelican. Many of our adult small gulls acquire black hoods in alternate plumage, and many of our adult terns acquire black caps; Black Tern actually becomes largely black.

As we've discussed, what would appear to be an alternate plumage is acquired partly, or even entirely (in Snow Bunting), as a result of the wearing of dull or pale feather tips. Often the behavior of a singing or displaying bird accentuates a part of its plumage that may have been less visible at other times of the year, as when Red-winged Blackbird thrusts its red "shoulders" (actually the lesser wing coverts near the "wrist" of the wing) outward when singing—thus its English name. At other times of the year, note the shoulder on Red-winged Blackbird in flight, but only a little red shows on the folded wing.

Not all seasonal changes involve the changing of a duller (though fresher) basic plumage to a more colorful alternate plumage. Feathers and the overall plumage are constantly exposed to sunlight and the elements (vegetation, dirt, and so on) and will inevitably wear and fade. By early August, our ground-dwelling sparrows often look extremely bedraggled in comparison to the fresh-plumaged ones seen in October. One striking example of this can be found in the last volume of Arthur Bent's well-known *Life Histories of North American Birds* (1968). The color frontispiece in this book—the third of three on cardinals, grosbeaks, buntings, towhees, finches, sparrows, and allies—is a painting by John A. Ruthven illustrating three recently described subspecies of Smith's Longspurs, a species previously considered to have no geographical variation. Joseph R. Jehl, Jr., revealed some years later that these differences were a result not of geographical variation but of wear and fading: The specimens on which the descriptions (and the painting) had been based were taken at progressively later dates in one summer from three different locations. Feather wear can also be seen in cavity nesters, especially woodpeckers, which often become very worn and stained, the results of a breeding season spent climbing in and out of tight tree cavities.

Wear and fading can give a species different appearances during the course of the year. These **Smith's Longspurs** from the frontispiece of the final volume of Bent's *Life Histories* were thought to represent three distinct subspecies. Later, noted ornithologist Joseph R. Jehl, Jr., observed that the differences were due to wear and fading. The more colorful top figures depict birds collected in spring; the paler bottom figures were from mid-summer, just before the prebasic molt.

Geographic Variation

One of the most interesting aspects of variation in birds is that which occurs among different geographical populations of the same species.

Evolution and speciation are ongoing processes, and the isolation and separation of populations have resulted in variation within a species. Such differences arise through genetic change in response to varying environments, as well as to the pressures of natural selection. If such differences are consistent within a certain geographical area, the geographical variants are formally recognized as distinct *subspecies*.

Subspecies are formally recognized by a trinomial for the scientific name: For example, in the scientific name for our Northern Harrier, *Circus cyaneus hudsonicus,* the third name designates the subspecies *(hudsonicus)*. Subspecies are sometimes also given an English name, often set off in quotes—such as "Harlan's Hawk," a northwestern subspecies of Red-tailed Hawk—but this English name isn't formerly designated by the Committee on Classification and Nomenclature of the American Ornithologists' Union. If no subspecies are recognized,

Unusual among eastern passerines, **Palm Warbler** has two very distinctive subspecies that differ in the timing of their migration and their distribution. The more numerous "Western Palm" (top) is a later spring and earlier fall migrant. In spring it is largely found west of the Appalachians. Most "Western Palms" winter in Florida and the West Indies. The more colorful "Yellow Palm" or "Eastern Palm" (bottom) migrates through the Atlantic states and does so earlier in spring and later in fall. It winters in the southeastern states, primarily from Louisiana to South Carolina. The two subspecies meet and interbreed to an unknown extent in central Quebec. (top: Alabama, April; bottom: Maine, June)

the species is considered *monotypic,* and only the scientific binomial is used. Thus the monotypic Common Loon is known as *Gavia immer.* But many species, including a majority of our passerines, are *polytypic;* that is, they have two or more named subspecies.

A simple example of this is Palm Warbler, which has two distinctive subspecies. "Western Palm Warbler" is the more numerous and widespread of the two subspecies. It winters mainly in Florida and the West Indies and migrates through the Ohio and Mississippi river valleys in spring, mainly during May. It breeds in the western two-thirds or so of the species' breeding range. It was described to science from its winter grounds on Hispaniola in 1789. "Yellow Palm Warbler," the other subspecies, was described by Robert Ridgway in 1876 from a Massachusetts specimen. It winters mainly in the southeastern United States and migrates up through the Atlantic states during April to its breeding grounds, from Maine through Atlantic Canada. It is both an earlier spring migrant and a later fall migrant than "Western Palm Warbler." Once a second subspecies is named, the first of the species described becomes the *nominate* race, in which the second and third names repeat. Thus, the full scientific name of "Western Palm Warbler" is *Dendroica palmarum palmarum* and "Yellow Palm Warbler" is *Dendroica palmarum hypochrysea.* Some species, like Palm Warbler, may have only a few named subspecies, but others have many. Yellow Warbler—with nearly 50 named subspecies—is an extreme example, though many of those are resident races found only in the Neotropics.

In North America, there tends to be much more geographical variation within species in the West. The West's multiple mountain ranges, valleys, and deserts act as both occupied habitat and habitat barriers; consequently, various populations are isolated from one another, a process leading to evolutionary change. The East is much more uniform, and geographic variation is slight. Palm Warbler is a notable exception within eastern birds; another is Short-billed Dowitcher, with two subspecies,

Willets have two distinctive subspecies that may actually be two full species. The smaller "Eastern Willet" *(left)* has a shorter bill with a pinkish base and is more extensively barred in breeding plumage. It nests in Atlantic and Gulf coast salt marshes and winters entirely south of North America. The plainer "Western Willet" *(right)* nests in freshwater marshes on the Great Plains and Great Basin and winters in coastal regions on both the Pacific and Atlantic sides of the continent. It is the only subspecies of Willet to be recorded as a migrant in the interior. *(left: Maine, May; right: Montana, June)*

Fox Sparrow shows complex geographical variation, and the four distinct subspecies groups may represent distinct species. Within each group, multiple subspecies have been described, though as with Song Sparrow, they have likely been over split. The **"Red"** northern group *(iliaca, top)* winters largely in the Southeast and is the only group that occurs (other than accidental occurrences) in eastern North America. Note the overall rufous coloration and the streaked back contrasting with a bright rufous tail. The bird shown is very bright—more typical of birds from the eastern part of the range. The **"Sooty"** Pacific coastal group *(unalaschcensis, bottom)* consists of seven named subspecies that get increasingly dark from north to south within their breeding range. The northern-breeding races are more migratory and winter farther south than the more southerly breeding ones. This individual represents one of the darkest races, probably *fuliginosa*. Note the overall dark coloration, the unstreaked back, and the browner, less rufescent tail. *(top: New Hampshire, November; bottom: British Columbia, February)*

griseus and *hendersoni,* in the East, which are distinctive from one another in alternate plumage. (A third, *caurinus,* is found on the Pacific coast.) The more strongly colored *hendersoni* has been confused with the similar Long-billed Dowitcher.

Willet is another shorebird with strong geographical variation. The Gulf and Atlantic coastal breeder, *semipalmata,* differs strongly in structure and plumage from the western interior breeder, *inornata.* The former subspecies completely vacates North America in fall, so Willets wintering on the Gulf and south Atlantic coasts—along with all interior migrants—are western *inornata.* Since there are also vocal differences, it remains unresolved whether these should be recognized as merely distinct subspecies or designated as distinct species. Marsh Wren, Nelson's Sharp-tailed Sparrow, and Seaside Sparrow also show

These maps illustrate the breeding *(top)* and winter *(bottom)* ranges of the four subspecies groups of **Fox Sparrow**. Birds from the White Mountains of eastern California and western Nevada are placed in the subspecies *canescens* of the "Slate-colored" group but have intermediate mitochondrial DNA. Most of these birds have call notes like the "Thick-billed" group farther west. The "Slate-colored" group migrates *southwest* to winter in California, one of the few western species to do so.

distinct geographical variation within each species in the East. But as in many other cases where two or more subspecies are described from the East, these subspecies are only weakly differentiated. More often in the West, subspecies are strongly differentiated, and additional genetic and other studies suggest that some of these subspecies should be recognized as full species.

What constitutes a subspecies is, of course, also a matter of debate. One standard, used for half a century or more, is that to be recognized as a distinct subspecies, 75 percent of the individuals of a population must be separable from 100 percent (or nearly so) of the individuals from another population. Some would prefer even a higher percentage, say, 90 or even 95 percent. A species that is considered monotypic might still show some geographical variation, but these differences are considered to be part of a *cline*, a gradual phenotypic change over a geographical gradient. Subspecies (as well as species)

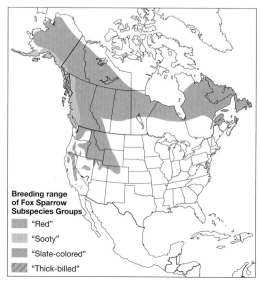

Breeding range of Fox Sparrow Subspecies Groups

- "Red"
- "Sooty"
- "Slate-colored"
- "Thick-billed"

boundaries are frequently a matter of debate. Many subspecies that were once described and named have since been *synonymized* (merged) with other subspecies. The American Ornithologists' Union's Committee on Classification and Nomenclature is officially charged with all aspects of taxonomy—including the recognition (or not) of subspecies—but its last published checklist that included subspecies was published in 1957 (*Check-list of North American Birds,* fifth edition). The next checklist (the eighth edition, expected after 2010) will again include subspecies.

Sometimes subspecies are placed in their own *group* of subspecies if they show similar characteristics that differentiate their group from at least one other group. These groups are informally designated with a scientific name—which is the first (chronologically) named subspecies within each group—and an English name. Fox Sparrow is an example of a polytypic species divided into subspecies groups that may be treated as full species in the future.

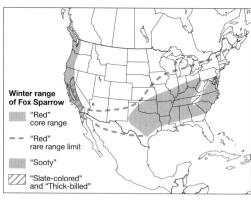

Winter range of Fox Sparrow

- "Red" core range
- - - "Red" rare range limit
- "Sooty"
- "Slate-colored" and "Thick-billed"

Fox Sparrow: A Complex Polytypic Species Fox Sparrow, in the monotypic genus *Passerella*, is a strongly polytypic species that has 19 named subspecies in four subspecies groups, described here:

■ **"Red Fox Sparrow"** The northern *iliaca* group comprises two poorly differentiated (from each other) subspecies. Birds from this group, the only group with a prominently streaked back, are strongly rufous overall. They are scarce in the West (except for those birds in the western part of the breeding range). Most of them winter in the East, making this the only subspecies group that occurs regularly in the eastern half of the continent.

■ **"Sooty Fox Sparrow"** This northwestern *unalaschcensis* group of dark brown Fox Sparrows comprises seven named subspecies breeding from the Alaska Peninsula to southern British Columbia. The southernmost subspecies are the darkest, the northern and westernmost the palest. Research done about a century ago documented that the northernmost subspecies migrate the farthest south, wintering south to northwestern Baja California, in a "leap frog" migration.

■ **"Slate-colored Fox Sparrow"** This western *schistacea* group of up to five named subspecies breeds in the northern and central Rockies and the Great Basin ranges, and it winters mainly in California. It and the next group are largely grayish and unstreaked above. Like the above groups, they have small bills.

■ **"Thick-billed Fox Sparrow"** This *megarhyncha* group from the Pacific states is composed of up to five subspecies that breed in the California mountain ranges, the southern and central Cascades of Oregon, and perhaps from southernmost Washington. It winters almost entirely in California and adjacent Baja California. Characteristic of this group is, as advertised, the thick bill (almost like that found on some of our thicker-billed grosbeaks) evident in the breeding subspecies in northwestern and southern California. The flank streaking is more sharply defined, on average, than in the "Slate-colored" group.

Fox Sparrows in the **"Slate-colored"** group (*schistacea, top*), breeding in the Rockies and Great Basin ranges, have small bills, rufous tails, and unstreaked gray backs. The similar **"Thick-billed"** group (*megarhyncha, bottom*) is thicker billed and more sharply streaked below. This "Thick-billed" represents one of the thickest billed subspecies, either *brevicauda* or *stephensi*. (*top: California, November; bottom: coastal California, November*)

Recent genetic work by Robert M. Zink and others indicates that there may be four separate species involved, although there is some genetic overlap in a narrow zone between the last two groups. The sweet and melodic songs of "Red Fox Sparrow" sound the most distinct from the other groups. The call notes of "Red Fox Sparrow," "Sooty Fox Sparrow," and "Slate-colored Fox Sparrow" all sound fairly similar, although experienced observers can recognize individuals from the "Slate-colored" group. However, the *chink* call of the "Thick-billed" group is totally unlike the others. Three of the four groups (excluding "Red Fox Sparrow") can be found wintering together in the mountains of southern California, but even there they divide into their own microhabitats.

The above cases are but a tiny sampling of North American species that exhibit distinct geographical variation, often with unresolved species-level ramifications. At least six additional cases involve emberizid sparrows: Spotted Towhee, Brewer's Sparrow, Sage Sparrow, Savannah Sparrow, White-crowned Sparrow, and Dark-eyed Junco. But even apart from species-level questions, learning about geographical variation and subspecies makes you more competent as a birder. Issues regarding subspecies are also becoming increasingly important in the public sphere, as ornithologists and conservationists attempt to preserve genetic diversity in an increasingly hostile environment.

Other Geographical Variation Beyond the official rules governing the recognition and naming of subspecies, there are some generalized "ecogeographic" rules that apply to geographical variation. These rules, named for the ecologists who first espoused them, are as follows:

These immature **Snow Geese** represent the two color morphs found within the subspecies known as "Lesser Snow Goose." The two morphs were once recognized as distinct species. The blue morph, known as "Blue Goose," is most numerous in the midsection of North America. (New Mexico, December)

■ **Allen's Rule** Within polytypic species, the extremities—such as the bill, legs, and (in mammals) ears—tend to be longer in warmer parts of the overall range and shorter in colder parts.

■ **Bergmann's Rule** Among polytypic species, overall size tends to be larger in cooler parts of the overall range and smaller in the warmer parts. This could be a north-to-south trend, although subspecies from even the southerly mountain ranges, where cool, may be large.

■ **Gloger's Rule** Among polytypic species, those from warm and humid areas are more heavily pigmented than those in cool and dry areas.

Other Variation

As you study the variations discussed below, keep in mind that birds are also individually variable. Birds of a species—even those of the same sex, subspecies, and age class—don't always look identical. This is known as *individual variation*. For instance, on a juvenile Broad-winged Hawk, the extent and strength of streaking on the underparts is variable. Young gulls during their various cycles prior to definitive adult plumage are notoriously variable, probably largely reflecting where they are in the molt process. And as we discussed in chapter 5, the outside elements (sun, heat, cold, rain, snow, amount of cloud cover, and the like) can change the appearance of an individual bird. Let us consider other well-known factors that make individual birds look different.

Color Morphs Certain species have *color morphs* that are familiar to birders. One such example is "Blue Goose" (formerly recognized as a separate species), a color morph that is numerous among the more easterly breeding populations of "Lesser Snow Goose" *(Anser caerulescens caerulescens)*. Color morphs can be common in populations, as in the above case, or rare, as in the blue-morph Ross's Goose. Dark-morph Broad-winged Hawks are very infrequent, but dark morphs are more regular (or even frequent) in other *Buteo* hawks such as Short-tailed, Swainson's, Red-tailed (western race *calurus*), Ferruginous, and Rough-legged. Sometimes a color morph is geographically restricted. For instance, white-morph Reddish Egret is infrequent at best in western Mexico but fairly common around the Gulf of Mexico and Caribbean. "Krider's" Red-tailed Hawk of the Great Plains is actually a color morph of the eastern subspecies *borealis* rather than a subspecies as recognized by the AOU (through 1957); it is designated as a morph because a majority of the birds from within its described range don't resemble these pale "Krider's," hence they are well short of the 75 percent rule used to recognize a subspecies as valid. Other species with distinct color morphs include Northern Fulmar, Red-footed Booby, Parasitic and Pomarine

Some species of gulls and terns acquire a pink flush to the white areas of the underparts. This is particularly frequent in Ross's and **Franklin's Gulls** (as in this breeding adult) and in Roseate and Elegant Terns. The coloration is caused by a carotenoid compound (astaxanthin) found within the feather. *(California, May)*

Jaegers, Eastern Screech-Owl, and White-throated Sparrow. Some birders refer incorrectly to the "white morph" of the Little Blue Heron. This white plumage is actually the first basic plumage. The same bird will look like the fully blue adult by the time it reaches 1¼ years of age.

Cosmetic Coloration and Staining

■ **Cosmetics** Occasionally when looking through a flock of gulls, you spot a pink-bodied bird. The rare Ross's Gull often has a strong pink flush, but so do many other small gulls. Laughing, Black-headed, Little, Bonaparte's, Sabine's, and especially Franklin's Gulls can acquire this color, and sometimes so do the medium-size gulls (e.g., Ringbilled). It is also routinely found on two species of terns, Elegant and Roseate (hence the latter's English name) and on adult male Common Mergansers. The pink flush is evident as a background color to the otherwise white underparts. Sometimes the wash is subtle; sometimes it is striking, particularly in lower light.

The causes of this pink coloration remained debatable for years. Some suggested the coloring was a substance from the preen gland that the bird groomed on its plumage. Finally, in 1990, after a study of Elegant Tern feathers, it was established that the coloration is caused by a common carotenoid compound (astaxanthin) found within the feather. Birds exhibiting this feature are said to have a "carotenoid flush."

■ **Staining** Waterbirds, particularly geese (such as Snow and Emperor Geese) occasionally get a rusty stain to their plumage, particularly visible on the white areas—much of the plumage on Snow Goose, the head and neck on Emperor Goose. Caused by water, vegetation, and mud, it is frequently noted among flocks of Sandhill Cranes, which habitually rub soil onto their plumage.

Hybrids Periodically during the course of a birding season, you run across a *hybrid*—an offspring whose parents are of two different species. (Some use *intergrade* to denote the crossing of two subspecies.) Cross-species mating is pretty routine among certain waterfowl (e.g., Mallard and American Black Duck), and just about any species within the same genus can hybridize. Within the waterfowl that show obvious male/female differences, it is the male that is characteristically identified. Hybrids occur regularly between Cinnamon and Blue-winged Teal and between Common and Barrow's Goldeneye, but it would be the brave soul indeed that tried to identify a female hybrid, when even pure birds look so similar to each other. Hybridization among certain large four-year gulls (Western, Glaucous-winged, and Herring) is also

Despite their strongly differentiated appearance, **Golden-winged** *(top left)* and **Blue-winged Warblers** *(top right)* are closely related and frequently hybridize. First-generation hybrids are known as **"Brewster's Warblers"** *(bottom left)*; recessive second- and later-generation hybrids are known as **"Lawrence's Warblers"** *(bottom right)*. These hybrids were described as separate species in the late 19th century. The birds shown are males. *(top left: Michigan, June; top right and bottom left: Connecticut, June; bottom right: Connecticut, May)*

When looking through flocks of waterfowl, it is not unusual to come across hybrids. Hybridization is frequent between some species, like Mallard and American Black Duck, and quite regular with others, like Blue-winged and Cinnamon Teal. It can occur with almost any waterfowl, particularly among captives in aviary collections. This striking male **Mallard x Northern Pintail** is more unusual. (California, April)

routine; one observer has referred to the these birds more than once as "avian weeds." Indeed, when reporting any rarity in these families, it is best to first consider the hybrid question: Could this "rarity" have been a hybrid? Other species groups where hybridization is frequent include the Yellow-bellied Sapsucker group, the Blue-winged and Golden-winged Warbler species pair, and the Townsend's and Hermit Warbler species pair. Hybrids between other species (e.g., Indigo and Lazuli Buntings, and Rose-breasted and Black-headed Grosbeaks) are regularly noted as well. Hybrids occur generally, but not always, between species of the same genus. While first-generation hybrids (known as *F1 hybrids*) typically look intermediate between the parent species, subsequent backcrossing (designated F2, F3, etc.) can render an appearance much more like one of the parent species. Hybrids have been discovered by examining the genetic code; plumage provided no outward sign of hybridization.

Some hybrids, like those in the Blue-winged x Golden-winged and Townsend's x Hermit Warbler species pairs, have been very well studied. First-generation hybrids show patterns of dominance, where one character of one of the species will dominate; others show incomplete dominance, so a character may be of one species or the other. A subsequent pairing of that individual with either one of the parent species or with another hybrid (hybrids are not at all necessarily sterile) can result in recessive genes manifesting themselves. In the Blue-winged x Golden-winged species pair, for instance, dominant-pattern hybrids are known as "Brewster's Warblers," and recessive individuals (second-generation or later hybrids) are known as "Lawrence's Warblers." These

This leucistic adult **White-crowned Sparrow** represents one of the coastal subspecies (*nuttalli* or *pugetensis*) with a dull yellow bill and more subdued crown pattern. While the extensive whitish feathering is at first glance startling, the head pattern readily reveals the species' identity. When encountering leucistic birds or true albinos, look for behavioral and structural traits that may offer identification clues; an additional clue may come from its normal-plumaged companions. (coastal California, November)

English names were given when these birds were described in the late 19th century as separate species. In the case of Blue-winged Warbler, this species appears to dominate over Golden-winged generally, and there is real concern about the long-term survival of the latter species. Similarly, Townsend's appears to dominate over Hermit Warbler.

Abnormal Plumages Occasionally, one sees an aberrantly colored individual. It may show irregular patches of white, with its other colors dull. If it's a passerine, don't assume it is Snow Bunting. It might be a *leucistic* individual of another species. Leucism results from the partial absence of pigment to the plumage. Complete absence of pigment results in a completely white or whitish plumage. A true albino will also have pink eyes and extremities (bill and legs). Leucistic or albinistic birds are believed to have lower fitness and survival probability. An increase in the amount of pigmentation, or *melanism,* causes birds with some black and brown plumage to present additional, uncommon black and brown coloring. *Xanthochroism* results from an abnormal dominance of yellow in the absence of the normal dark colors. Other color mutations result as well. One rare aberrancy is *gynandromorphism,* in which an individual is half-male and half-female—usually, one side is malelike and the other is femalelike. This aberrancy has been noted in a number of passerines (such as Black-throated Blue Warbler), in Ring-necked Pheasant, and in Northern Flicker.

This immature fall **Rose-breasted Grosbeak** is a very rare bilateral gynandromorph—an individual that exhibits male features on one side and female features on the other. Note the pink malelike wing linings on one side and the yellow femalelike wing linings on the other side. (Pennsylvania, September)

CHAPTER 7

IDENTIFICATION CHALLENGES

W hen you start birding, *every* species can present identification challenges. Each time you learn to correctly identify a new species—whether Northern Cardinal or Blue Jay in the East, or White-crowned Sparrow or House Finch in the West—is a confidence-building experience. The process of learning identification skills and becoming competent in the field can be frustrating and, as in any pursuit, some people learn more quickly than others. You will make mistakes periodically, but rather than being discouraged, regularly ask yourself whether you are more competent in the field than you were a year ago—or, for newer birders, a month ago. Assuming that you have been fairly active and that birding is a serious interest, the answer to that question is bound to be yes!

In this chapter we will give examples of North American species that we consider fairly easy to identify and some that are moderately difficult to tell. We will conclude with a list of what we consider the dozen most difficult identification problems in North America. Note that any species can be—and has been—confused with another. We recall hearing the story of a novice birder in southern Ontario who called an experienced birding friend one spring to report a disabled Common Loon in their yard. Since the brief verbal description sounded correct, the experienced birder suggested filling the bathtub and putting the bird there until it could be secured for rehabilitation at a center. Shortly thereafter, the novice called back and said the "loon" was screaming and was trying to get out of the tub. The experienced birder drove straight over and found a poor, drowned Hairy Woodpecker in the tub!

The two species of dark *Plegadis* ibis present an identification challenge. Breeding adults, such as this **White-faced Ibis** *(right)*, have a distinct face pattern; immatures *(left)* are more problematical. Opposite, two male **Black Rosy-Finches** *(top and bottom)* join two **Brown-capped Rosy-Finches**—a male *(second from bottom)* and a female *(second from top)*—at a Rocky Mountain feeder in winter. *(above: California, May; opposite: New Mexico, December)*

Easy Identifications

Keeping the above caveat always in mind, let's look at a few examples of easy identifications.

Egrets Great Egret is the widely recognized symbol of the National Audubon Society. It reflected the early years of the society's history, when plume hunters (acquiring feathers for ladies' hats) had severely depleted Great Egret populations. It's not likely, however, that most nonbirders know there is more than one species of egret.

In many of our wetlands two species occur: Great Egret and Snowy Egret. They can be easily sorted out by size and by bill and foot color. They also differ in their feeding behavior, with Snowy Egret being much more active. In some parts of North America, one can find another species of white egret, Cattle Egret. Looking like a pint-size Great Egret with a stubby bill and shorter neck and legs, it generally shuns wetlands, except for nesting, preferring fields; and, as its English name suggests, it often associates with domestic cattle. There are, of course, other white herons to consider in some areas. These include the white-morph Reddish Egret, the "Great White Heron" (currently considered part of the Great Blue Heron species), and the immature Little Blue Heron. The latter species, Little Blue, could be deemed moderately difficult to separate from Snowy Egret; to distinguish it, pay careful attention to the color of the bill, lores, legs, and feet, and notice the overall behavior. While nonbirders and beginners think of egrets as white, there is one species (Reddish Egret) in which the dark morph predominates. And, of course, the term *egret* is used collectively for a group of species (in North America represented by three different genera) that aren't particularly closely related. Beginners using older, hand-me-down field guides may be confused by the various English names used for one species, Great Egret. In earlier times it was known as American Egret or Common Egret. If at some point the New World subspecies of Great Egret *(egretta)* is split from the Old World subspecies, one of those older names may be used again.

American Coots and Common Moorhens With these birds, the first challenge will be realizing that even though they may swim like ducks, these birds aren't ducks. Nor do they sound like ducks. They are in a large worldwide family, comprising nearly 150 species, that includes rails. Coots are easily separated from moorhens and gallinules by their white bill. They also have conspicuously lobed toes. Coots are certainly more readily visible and often come for handouts, whereas moorhens and gallinules are more skulking and retiring. When swimming, moorhens and gallinules sink their chests deeper into the water and raise their tails above the water level more than coots do. Coots are overall more blackish. In certain parts of North America, we find Purple Gallinule, a very colorful species (although juveniles are a dull buffy brown). As a vagrant,

Purple Gallinule has occurred nearly continent-wide. It can be easily differentiated from both American Coot and Common Moorhen by its coloration and by its very long yellow legs and toes.

Caspian and Royal Terns Separating terns from gulls will be among the challenges a new birder faces. Clues to identification include the terns' long slender bills and disproportionately longer, more pointed wings. Many species of terns are strikingly white and have forked tails. Unlike gulls, most terns plunge-dive into the water. Along the Gulf Coast, the Atlantic coast, and the Pacific coast of southern California, two large red- or orange-billed terns often occur together, Royal Tern and Caspian Tern; the latter is actually the largest tern in the world. They can be rather easily separated by bill color and shape: Royal has a thinner orange bill. Caspian's blackish coloring on the inner webs of the primaries is easily visible as a dark patch on the underside of the wings; and in no plumage does Caspian have a pure white forehead like that exhibited by the Royal Tern below. The Royal pictured is an adult that has already lost its black cap, which is held only briefly in this species. A third, somewhat smaller solid-orange-billed tern, Elegant Tern, can be found with Royals and Caspians on the coast of southern California. Separating Elegants from Royals is at least moderately difficult. Elegant's bill appears more slender and longer—actual culmen length is about equal in the two species, but Elegant's bill is *relatively* longer— and in nonbreeding plumages it averages more black around the eye.

These three medium- to large-size terns represent a fairly straightforward identification. This **Royal Tern** *(right)* has already lost its black cap and has a white forehead, while the larger **Caspian Tern** *(center)* still has its black cap. Caspians have a thick blood red bill with a black tip and never show a pure white forehead. (They're streaked in winter.) The smaller **Sandwich Tern** *(left)* has a long black bill with a yellow tip. *(Texas, June)*

The calls of all three species differ, the most distinct being Caspian's, which varies from the raspy-sounding adult calls to the whistled calls of the young up to a year old *(see photograph, page 123)*.

Moderately Difficult Identifications

The following are moderately difficult identifications.

Ross's and Snow Geese The populations of white geese *(see photograph, page 91)* have greatly increased in recent decades, so much so that an additional hunting season has been added to cull their numbers. Ross's Goose was at one time thought to be a rather scarce species, known primarily in winter from California. Growing populations and increased detection have now shown it to be regular in many parts of North America and even common in parts of the West. It has occurred (at least as a rarity) continent-wide. Ross's and Snow can be easily distinguished by size and structure; Ross's stubby bill, which lacks black "lips," is particularly useful for identification. Unlike Snow Goose, the immature Ross's closely resembles the adult. Another fact to consider is that although a dark-morph Ross's is very rare, a dark-morph Snow is common, even abundant, within some populations.

Female Mergansers The three North American mergansers present a variety of identification problems. (A fourth, Smew, is a rare or casual visitor from Eurasia.) With mergansers, the immature male closely resembles the female until well through winter. Hooded Merganser—with its smaller size, its darker head with a blond rear crest, and its yellow-based bill—is relatively straightforward. Red-breasted and Common Mergansers *(opposite)*, however, have the same general coloration with rufous crests. Compared to Red-breasted, Common is distinctly larger, and the North American subspecies *(americanus)* has a much thicker-based bill. Overall, Common shows more contrast.

Winter Loons These are inevitably tricky, especially when viewed from a distance. While there are size differences among winter loons, identifying a single bird can be challenging. Away from the coast, Common Loon is always the expected species, though the others have all occurred, most at least casually. Look carefully at the winter loon's head and neck pattern and its bill shape and size. With Yellow-billed Loon, bill color and overall coloration (pale and brownish) are critical marks. Viewed in flight, the frequency of loons' wingbeats can prove helpful for identification: Common and Yellow-billed Loons have a slower flapping rate. These two species also often fly higher and extend the webs

of their feet out in flight, thus appearing to have bigger feet. Other differentiating characteristics are that Red-throated often dips its neck down below the horizontal level when flying and that Pacific is the only species that routinely migrates in large flocks.

Cormorants Except for those cormorants found in a few states along and near the Mexican border, Double-crested Cormorant is the only cormorant that is regular away from the coasts. Along the West Coast south of Alaska, three cormorant species occur; on the Atlantic coast two are found, though Great Cormorant is primarily a winter visitor south of the Maritime Provinces. Important characteristics for identification include the shape and coloration of the bare skin around the face. When viewing birds in flight, look at whether the neck is distinctly kinked; it will be held out straight in Pelagic and Brandt's Cormorants. Double-crested Cormorant, which often flies high, has particularly long, pointed wings. Relative tail length is a critical feature in separating Neotropic from Double-crested as well as Pelagic from Brandt's.

Yellowlegs Size is a good distinction between Greater and Lesser Yellowlegs, but identifying single birds can be tricky. In addition to noting size, look also at structure; the short bill of Lesser is a particularly useful feature. With the exception of alternate-plumaged adults, bill color—which is darker on Lesser—is a good distinguishing characteristic. Their feeding behavior is an underappreciated differentiating feature between

This remarkable flight shot of two female-plumaged mergansers shows both a **Common** (left) and a **Red-breasted Merganser** (right). The larger Common has a darker rufous head that contrasts sharply with its white chin and pale breast. These features are more blended in Red-breasted. Note also the thicker-based bill of Common, a characteristic of the North American sub-species, *americanus*. (Ontario, November)

these two species: Greater is much more active, dashing about in pursuit of food. Calls are also always diagnostic, and Greater is much more widespread as a winter visitor in North America.

Pewees and "Empids" (*Empidonax* Flycatchers) Intermediate birders often want to dive right into identifying the various *Empidonax* species without first mastering the other look-alike flycatcher groups—the phoebes (genus *Sayornis*), which dip and spread their tails downward, and particularly the pewees (genus *Contopus*), which don't bob their tails at all. The empids habitually flick their tail up, the exceptions being males singing on territory and Gray Flycatcher, which bobs its tail downward, like phoebes. The empids have relatively long tails and short wings; in contrast, pewees, including the Olive-sided Flycatcher, have short tails and long wings. Compare the photographs of Western Wood-Pewee and Least Flycatcher below, and notice in particular the long primary projection past the tertials on Western Wood-Pewee. In the West, Western Wood-Pewees are routinely confused with the western subspecies of Willow Flycatcher.

The first step in distinguishing among flycatchers is sorting them into the correct genus. Compare the long primary projection on the **Western Wood-Pewee** *(left)*, in the genus *Contopus*, to the much shorter primary projection on the **Least Flycatcher** *(right)*, one of the *Empidonax*. There are additional important behavioral clues: Most notably, *Empidonax* flick their tails after landing, whereas *Contopus* usually sit as still as a statue. *(left: California, June; right: Maine, June)*

Brown Swallows This challenge involves separating not only Bank from Northern Rough-winged Swallows, but also Banks from young Tree Swallows, which, like Banks, are also largely brownish above and can show a banded effect across the breast. Note, however, that on Bank the pale of the throat wraps around the sides of the neck. Bank, our

long primary extension past tertials

short

long tail

smallest swallow, has rapid wingbeats; Northern Rough-winged, which is one of our largest swallows, has slower, floppy wingbeats.

Fall Warblers Some fall warblers can be deemed moderately difficult to tell apart, but contrary to popular belief, in most species, there is little difference in their appearance between spring and fall. However, some of the most abundant species (e.g., especially in the genus *Dendroica*) do show significant differences between spring and fall. Some identifications, such as separating Bay-breasted and Blackpoll, are hard. And even experts can find distinguishing between Mourning and MacGillivray's Warblers extremely difficult *(see page 169)*.

House and Purple Finches House Finch is now the most common finch species at most feeders, particularly in urban areas. Adult male House and Purple Finches differ from one another in color; a more appropriate name for Purple Finch would perhaps be "raspberry finch." Females and immature males—which hold a femalelike plumage for over a year in Purple but just a few weeks in House—can be easily separated by the face pattern: Purple's pale supercilium contrasts with a dark cheek. The Pacific subspecies of Purple *(californicus)* is buffier below and has more diffuse streaking than the nominate eastern subspecies.

Difficult Identifications
Most of the following examples present problems for birders of all experience levels.

One normally thinks of separating **Bank Swallow** *(left)* from Northern Rough-winged *(see photograph, page 29)* when differentiating the brown swallows, but immature **Tree Swallow** *(right)* is also largely brownish and can show a bit of a breast band. Note on Bank how the pale of the throat wraps around the darker ear coverts. Although not apparent here, Bank is our smallest swallow and has a rapid fluttery flight. Knowing the flight styles of the various swallows is one of the best ways to initially separate them into species with the naked eye. *(left: California, October; right: California, August)*

Trumpeter and Tundra Swans All swans look large, though when these two species are together, Trumpeters appear huge. In adults, look carefully at the way the black skin in the anterior portion of the face meets the eye and cuts across the forehead: In Tundra it cuts straight across and closes in front of the eye; in Trumpeter it dips into a V and is inclusive of the eye. Also listen for any diagnostic vocalizations. Telling the young apart, other than by size, qualifies as one of our most difficult identifications; one clue is that Trumpeters stay browner longer into winter.

Female-Type Blue-winged and Cinnamon Teal Overall coloration and pattern, especially the face pattern, are particularly important in separating these two species. Blue-winged is grayer and more

It is always a challenge to separate female-plumaged teal, particularly **Blue-winged** (top) and **Cinnamon Teal** (bottom). Both have streaked undertail coverts that are uniform with the rest of the underparts, unlike the smaller Green-winged Teal. Blue-winged's coloration is overall more mottled and colder. It also has a smaller bill, paler lores, a stronger eye ring, and a more contrasty eye line. The female Cinnamon—sexed by iris color and the lack of a broad white border behind the blue forewing, in addition to overall coloration—is buffier overall and larger billed, and it has a less patterned face. (top: Florida, March; bottom: Arizona, April)

patterned overall; it has a stronger eye line as well as a paler lore region with a bolder eye ring. Cinnamon's larger bill size is an excellent distinguishing characteristic. Fall male Blue-wingeds resemble females, but they can be separated from females in flight by the wing pattern. Immature male Cinnamons look like female Cinnamons, apart from wing pattern, and thus like Blue-wingeds; so it is important to carefully note the iris color: A reddish iris indicates a male Cinnamon. Range can sometimes be helpful in separating these species because Cinnamon occurs only casually in eastern North America. Cinnamon and Blue-winged can be separated from the smaller, small-billed Green-winged by their uniform, rather than whitish, undertail coverts.

buffy

Greater and Lesser Scaup Generally, if you have a Greater Scaup you know it. Identification issues come with those Lesser Scaup that appear to have smooth, round heads. This appearance is found particularly among diving birds, when Lesser Scaup's peaked rear crown and "notch" largely disappear *(see pages 65 and 111)*. When viewing mixed flocks, keep in mind that Greater Scaup is the larger species. If you think you've found a Greater

white collar

among a flock of Lessers and this Greater seems to be the same size as the other birds, your identification may be wrong. Check the head, which on a Greater really does appear round. Other diagnostic features, like the extent of the dark tip on the bill, can be observed at close range.

Glossy and White-faced Ibis With Glossy Ibis spreading west and records of White-faced Ibis increasing in the East, identification of these species is now an issue continent-wide. To distinguish one from the other, look very critically at the color and pattern of the face, including the skin in front of the eye. All adult (and even most juvenile) Glossy Ibis will show a thin powder-blue line above and below the facial skin but not around the eye. Adult White-faced Ibis can always be identified by its reddish iris color and often by pinkish skin in the lores. A young White-faced Ibis that lacks any pattern in the face may not be separable from some young Glossy Ibis. By early spring, though *(see page 154)*, White-faced Ibis will have pinkish bare skin in the lores and at least some reddish tint to the iris.

Separating female and immature hummingbirds requires care. Some separations, such as **Rufous Hummingbird** *(bottom)* from Allen's, are nearly impossible in the field. Separating female and immature **Broad-tails** *(top)* from these is also very difficult. Broad-tailed appears longer tailed and has more-blended buffy underparts. Rufous/Allen's types have a white-collared appearance. The chip notes are utterly distinctive, once learned. The smaller and differently structured Calliope (very short tailed) is quite similar in coloration to Broad-tailed. *(top: Arizona, April; bottom: California, August)*

Accipiters: Sharp-shinned Hawks, Cooper's Hawks, and Northern Goshawks Much has been written on the subject of separating accipiters, and it is beyond the scope of this coverage to offer anything definitive here. Given the difficulties in differentiating between *Accipiter* species, there are always birds that will be recorded in your journal as "*Accipiter* (species uncertain)." The Cooper's Hawk population has substantially increased in recent decades, which has made distinguishing between Sharp-shinned Hawk and Cooper's Hawk particularly challenging. In addition, we believe that reports of Northern Goshawks are inflated because of its similarity to female Cooper's Hawk. For Northern Goshawk, identification is pretty straightforward. One underappreciated feature of Northern Goshawk is how long the wings are; hence it gives quite a different flight silhouette than Cooper's gives.

Sterna Terns: Roseate, Common, Arctic, and Forster's Terns When identifying adult *Sterna* terns, consider both the time of year and the extent of molt. As noted in chapter 6 *(see photograph, page 135)*, adult Common and Arctic Terns retain their black caps into fall, whereas Forster's and Roseate Terns lose their caps. Concerning the immatures, the pattern of the lesser coverts is pale on Forster's, dark in the others—an important clue. As with other terns, calls are important, as is range: Roseate Tern is virtually unknown in the Gulf of Mexico (away from the Dry Tortugas region) and inland.

Jaegers Nearly every pelagic birder can relate stories of arguments about whether a jaeger was one species or another that were followed by definitive evidence (such as a photograph) revealing that the bird was a third, unconsidered species! Here are three things to remember when you're identifying jaegers: First, juvenile Pomarine Jaegers normally arrive very late during the fall season, unlike juveniles of the other two species—Parasitic and Long-tailed—which arrive by late August. Second, away from the Great Lakes and the Salton Sea, Long-tailed Jaeger is just as likely as the others (if not more likely) to appear inland. Third, be careful not to confuse dark-morph Pomarines lacking their long, central, extended tail feathers with young South Polar Skuas.

***Selasphorus* Hummingbirds: Broad-tailed, Rufous, and Allen's Hummingbirds** The difficulties of separating female and immature male Rufous and Allen's Hummingbirds are well known, but even adult males can be tricky because some Rufous Hummingbirds have green backs. A hummingbird with a rufous back is, of course, a male Rufous; this color is acquired by many immature males during early winter. An

whitish primary coverts

dusky primary coverts

pale primaries due to reflected sunlight

underappreciated problem, however, is separating a female-type Broad-tailed Hummingbird from Rufous and Allen's *(see photographs, page 163)*. Although Broad-tailed occurs annually in the Southeast, it is almost accidental along the West Coast. The buff on the underparts of Broad-tailed is slightly duller than that found on Rufous and Allen's, but it is more uniformly and extensively distributed, including to the sides of the neck. Once recognized, Broad-tailed's more metallic *chip* call notes are distinctive. Female-type diminutive Calliopes also closely resemble female-type Broad-taileds in coloration and pattern, though not in structure, and have been misidentified by experienced birders.

Juniper and Oak Titmice Juniper and Oak Titmice were recently split from a single species, the Plain Titmouse. These look exceedingly similar, but apart from a few narrow regions in northern and eastern California, they do not occur together. Despite their nearly identical appearance, they sound quite distinct; for example, the chattering calls of Juniper Titmouse closely resemble calls of Bridled Titmouse.

Curve-billed and Bendire's Thrashers While fresh-plumaged fall birds can be told using a suite of field marks—including Curve-billed's longer, heavier bill—worn summer birds are more difficult, especially because shorter-billed juvenile Curve-billed Thrashers are about. Curve-billed is more talkative in general, and its loud whistled calls are distinctive. Apart from singing, Bendire's is largely silent. Curve-billed shows distinct geographical variation (including in its call notes).

Distinguishing **Sooty Shearwater** *(left)* from **Short-tailed Shearwater** *(right)* represents one of the toughest identification challenges for pelagic observers. The larger Sooty has a longer and thicker bill; Sooty also usually has a paler underwing and whitish underwing primary coverts. Short-tailed averages darker and more uniform on the underwing, but this is a light-dependent field mark and often hard to assess. Note the feet projecting beyond the tail on the Sooty—mistakenly thought by some to be a useful field mark for Short-tailed. *(left: California, September; right: California, November)*

Fall Chipping and Clay-colored Sparrows In eastern North America, Chipping Sparrow predominates, but Clay-colored does regularly occur. The same situation applies in the West, although here the similar Brewer's Sparrow gets thrown into the mix. Certainly Clay-colored is overall the buffiest and has the most-patterned head. When separating these three species, look carefully at the overall head pattern—particularly the color of the lores and the presence (or absence) of a distinct median crown stripe or a moustachial stripe *(see pages 46-47)*.

The Hardest Dozen

It goes almost without saying that major progress has been made in field identification since the first field guides appeared in the early 1900s, so when we say that certain species are not field identifiable or are separated only under the best of circumstances, we should always add the disclaimer *based on present knowledge*. Some species pairs considered difficult or nearly impossible to separate several decades ago can now be separated with relative ease. These major advances in field identification have resulted in part from better optics, but progress can also be attributed to the number of talented individuals critically looking at and addressing these problems. The advances stem as well from greater international communication among field-identification experts.

The list below represents 12 of the most difficult identification problems, given here in taxonomic order: (1) Sooty and Short-tailed Shearwaters, (2) basic-plumaged dowitchers, (3) Chimney and Vaux's Swifts, (4) female-type Ruby-throated and Black-chinned Hummingbirds, (5) female-type Rufous and Allen's Hummingbirds, (6) Eastern and Western Wood-Pewees, (7) Hammond's and Dusky Flycatchers, (8) Carolina and Black-capped Chickadees, (9) Gray-cheeked and Bicknell's Thrushes, (10) immature Mourning and MacGillivray's Warblers, (11) Eastern and Western Meadowlarks, and (12) Common and Hoary Redpolls. Below, you'll find some brief information on five of these challenging identifications. Specialty books on individual families, such as hummingbirds, will have detailed information on many of these field problems, and magazines often have relevant articles as well *(see page 201)*.

You may notice that our list does not include American Crow and Northwestern Crow. They have been excluded because they are impossible to separate with certainty in the field, other than by range. In areas where they overlap (such as western Washington), they are alleged to hybridize extensively, and in those areas crows aren't identified to species. Whether they continue to be treated as separate species remains a matter of debate. It should be added that separating Fish from American

browner, darker, and more uniform breast

grayer breast with obscure spotting

Separating dowitchers in any plumage requires careful attention, but in basic (winter) plumage it is extremely difficult. **Long-billed Dowitcher** *(top)* is overall a little darker and browner and has slightly longer legs; **Short-billed** *(bottom)* is paler, grayer, and shorter legged. Short-billed's breast appears slightly spotted; this area is more uniform on the Long-billed. Location and habitat are separating features in most instances, and the calls are diagnostic. *(top: California, January; bottom: Texas, September)*

Crows is also very difficult, as is telling Chihuahuan from Common Ravens. Although there are some structural and plumage features that separate Fish from American Crow *(see photographs, page 120)*, as indicated, voice is the best distinction. Size separates Chihuahuan from Common Raven, but assessing a single bird on that feature alone is always difficult. Chihuahuan has a shorter bill (appearing disproportionately thicker, with shorter projection past the bristles) and a shorter tail.

We have also omitted separating Cordilleran from Pacific-slope Flycatcher, which is impossible in the field based on today's knowledge. These species cannot be identified other than by song and, especially, male "position" call notes (given year-round). These two were formerly considered one species, Western Flycatcher, and many seen in the field are prudently still called that; some dispute this split.

Sooty and Short-tailed Shearwaters Sooty and Short-tailed Shearwaters are overall dark and medium in size *(see photographs, page 165)*. Because Sooty occurs in both the Atlantic and the Pacific oceans but Short-tailed is restricted to the Pacific, separating them is a West Coast problem; however, there is a specimen record of Short-tailed for Florida and a sighting off Virginia. There are a few seabird experts in the West who feel comfortable separating these two species, based partly on their flight behavior. Visually, one of the characteristic features of Short-tailed is the more petite bill, as the old English name, Slender-billed Shearwater, indicated. Although Short-tailed Shearwaters are usually duskier under the wings, both species are individually variable—to say nothing about viewing birds under differing light conditions. More than once we have watched experienced observers identify a distant bird with certainty to species, only to have them become less certain when the bird came right behind the boat. And on occasion we have heard these individuals say, "They're easier to identify when they're farther away!" It's a bit of a cop-out to say so, but in the hand the measurements are diagnostic.

Basic-Plumaged Dowitchers It wasn't until the middle of the 20th century that the two dowitchers, Short-billed and Long-billed *(see photographs, page 167)*, were recognized as separate species. Significant contributions on separating them have been made over the last few decades. Their alternate and juvenal plumages have now been pretty well worked out, but basic-plumaged birds remain problematic, unless vocalizations are heard. Short-billed is somewhat paler and grayer overall (and its tail is paler on average) with more distinct breast spotting; general structural clues are also helpful, if not diagnostic. Habitat differences can also be used, and any inland dowitcher in winter is almost certainly a Long-billed.

In eastern North America there are two very similar chickadees: **Black-capped** *(left)* and **Carolina Chickadee** *(right)*. While range usually separates the two, there is a narrow overlap zone where the two hybridize. In addition, Black-capped periodically pushes south in fall and winter and appears from eastern Kentucky to the mid-Atlantic piedmont, well within Carolina's range. On Black-capped Chickadee, note the whiter rear auricular (washed with gray on Carolina) and the bolder white edges on the greater coverts and secondaries (concealed on this Carolina). Black-capped is slightly warmer colored on the back and flanks (grayer in Carolina). Vocalizations are important separating features too. *(left: Minnesota, February; right: Ohio, October)*

uniformly white

more prominent whitish edges

shaded gray at rear

Carolina and Black-capped Chickadees In fresh plumage, Carolina and Black-capped Chickadees appear very similar, differing only slightly in cheek color and wing pattern. Both species show geographical variation. While range largely separates them, Black-capped does periodically stage fall and winter incursions south into Carolina's range in the East (eastern Kentucky to mid-Atlantic states). Vocalizations are useful, but there is a narrow, well-studied zone of hybridization where it is best to leave most chickadees unidentified.

Immature Mourning and MacGillivray's Warblers As we discussed earlier, most warblers, even in fall, are not difficult to tell apart. This is undoubtedly the most difficult wood-warbler identification problem. Keep in mind that Connecticut, of the same genus, could be included here, though it differs significantly in its behavior: It's a walking warbler. Mourning and MacGillivray's Warblers have largely different ranges, but the former species occurs rarely to casually in the West and MacGillivray's is casual in the East. Mournings have, on average, a thinner yet more complete eye ring, a shorter tail, and a yellower throat, which is usually pale gray on MacGillivray's. There seems to be overlap on all of these points, so call notes are the best distinguishing feature.

Common and Hoary Redpolls Although the frosty pale extremes of Hoary Redpoll are distinctive—Commons are darker—the problem in identification arises among at least some of the younger and overall slightly darker and more streaked Hoaries: Are these dark Hoaries, or very pale Commons, or hybrids (although records of hybridization are not well documented)? If there are any vocal differences, they have yet to be well articulated; clearly much more study is needed. Both species show geographic variation, so it's best to remember that the vast majority of redpolls in southern Canada and the U.S. are Commons.

Separating the *Oporornis* warblers represents the most difficult task within the wood-warbler family. This is particularly the case with nonadult males of **Mourning Warbler** *(left)* and **MacGillivray's Warbler** *(right)*. Range is useful, but both species occur as vagrants in the range of the other. MacGillivray's is slightly longer tailed, has a more obviously broken eye ring with thick whitish eye arcs, and a grayer throat. Note how the yellow extends into the throat region on this immature male Mourning—sexed by the slight blackish stippling on the sides of its throat, which is diagnostic for Mourning. Some birds remain problematical and are best separated by their diagnostic chip notes. *(left: New Jersey, October; right: California, August)*

CHAPTER 8

FIELDCRAFT

Fieldcraft skills—how to search for birds, how to move, and how to communicate what you see to others—can be as important as your knowledge of bird identification. Understanding the basic concepts of fieldcraft gets you into position to see more birds, and your skills, honed by this experience, will continue to develop. This chapter, which builds on chapter 2, "How to Get Started," is a potpourri of how-to skills and information—all of it essential to becoming a well-rounded birder.

Most birders start out with a local focus but soon become fascinated with birding new locations. Becoming familiar with the where and when of birding, regionally or continent-wide, will keep you attuned to both the wider natural world and the fraternity of birders, so we'll discuss birding at these different levels. It will also help to understand some of the life history of the birds that you see—whether they are migrating, nesting, wintering, or truly rare—and to know the best ways to keep track of and document your sightings. At the end of the chapter, we'll delve into the essential books and journals for birders and discuss some additional equipment.

To see a **Colima Warbler** (male, *above*) in the United States, you'll need to make a trip to the Chisos Mountains of Big Bend National Park in Texas. A ten-mile round-trip hike to Boot Spring (*opposite*, notice the boot-shaped rock formation) will take you into Colima's nesting habitat of coniferous and oak woodland with grassy ground cover. (*above: Texas, May*)

Observing Birds

Many of us grew up with little chance to become familiar with the natural movements and instincts of wild animals. All the accoutrements of modern life have insulated us from the natural world. Yet as birders, we make a conscious effort to connect with nature. You'll soon wonder how nonbirders around you can hear and see so few birds. In fact, most veteran birders are *unable* to ignore birds; simply hearing a distant song through an open window or seeing a flicker of movement makes an immediate impression, sometimes at inopportune moments. Some part of your mind is always trolling for clues and is ready to engage when random bird events present themselves.

When you're actively birding in a natural environment, finding birds can run the gamut from obvious (open situations with large birds), to difficult (warblers in the tops of tall, leafy trees), to nearly impossible (dense marshes with secretive rails). Although several variables—such as lighting, birds' activity level, and accessibility of sites—influence conditions, there are ways to optimize your chances of finding birds.

First Steps When you arrive at a new birding spot, recall the advice you got when you learned to cross the street: Stop, look, and listen. Put yourself into position to see and hear any bird activity before you move around too much. Birds of interest might be close by. They might be hiding from you, or they may be secretive and elusive by nature. Give them a few minutes to return to their natural activity level before you start moving. In good locations you might not get beyond the parking lot for half an hour. When you do move, move slowly.

Using Your Ears Wherever you are birding, make a conscious effort to listen for bird sounds. Beginning birders often focus on the sounds of songbirds and ignore the equally distinctive calls of hawks, shorebirds, gulls, and a host of other species. You don't have to recognize the song or call for it to be useful. Often a sound will be your first clue to *where* a bird is. Let your ears lead your eyes instead of walking about at random.

Learning songs and call notes takes time; the repeated experience of seeing, identifying, and watching a vocalizing bird connects the sound with the image and the name. We discussed techniques to speed up the learning process in chapter 5 *(see page 86)*. Any additional effort you put into memorizing songs and calls will greatly increase your ability to find and identify birds.

Conditions can be unfavorable. Many vocalizations are soft or are given by distant birds, so you need to be as quiet as possible. Take your clothing into account: Avoid noisy synthetics, such as nylon windbreakers. Wind and traffic noise also obscure bird sounds; under those circumstances, alter your search tactics or location. Cupping your ears with both hands and slowly swiveling your head pulls in distant sounds and also gives you a directional clue. You may find that some sounds have a ventriloquial quality and are exceedingly difficult to pinpoint. If the bird is a persistent vocalizer, attempt to walk toward the sound and scan for movement. The singer might be behind a branch or leaf, so changing your location might reveal its location.

Other types of audible clues include the concerted alarm calls of nearby birds, often used to signal the presence of a predator. When you hear this behavior, check the area for a perched or flying raptor. The rustling of leaves is another clue, often revealing the presence of ground-foraging birds such as towhees and thrashers.

Using Your Eyes Some visual skills that birders acquire may not be obvious to the beginner. We covered the basics of using binoculars in chapter 2, but a few points are worth repeating. For instance, find birds with the naked eye, not your binoculars. The binoculars' field of view is too restricted for general scanning. In open habitat, scanning with binoculars is sometimes useful—to check pond edges, a distant ridge-line for migrating raptors, or the ocean's horizon for distant seabirds—but the technique rarely yields results in forests and fields.

Looking for movement is a very effective way of finding birds. Movement in a natural setting often means the presence of a bird, and birders develop the ability to tune out movements not associated with birds. To look for movement, *you* need to stand still. Take in the wide picture rather than concentrating on a specific area, and move your head slowly. Wind is distracting when looking for birds. The continuous motion of leaves and other vegetation masks the movements of birds, making them very difficult to locate. This is a good reason to bird early in the morning, when the wind is often mild.

Unless the day is overcast, most birders prefer to wear a hat to shade their eyes. Strong light and glare cause the pupils to contract, reducing the ability to see subtle details. If you don't have a hat, position yourself in a shaded area. For example, if you want to look into a tree

Vocalizations aren't the only clues to a hidden bird's presence. **Brown Thrasher** often reveals its presence by the sounds it makes scratching in the leaf litter. When birding, stay alert for other nonvocal sounds, such as the wing trill of a hummingbird or the drumming of a woodpecker. *(New Jersey, November)*

that is in line with the sun, use the shadow of a nearby tree trunk to block the direct sun from your view.

Getting Close You don't need to get close to birds to identify them, but it often helps. Use caution, though, for abrupt movements frighten birds, who will instinctively flee or hide. Even raising your binoculars can startle nearby birds, so move slowly and quietly.

Birds have become habituated to people in some locations, especially where they are fed; bird feeders, city parks, and boat docks are good examples. Some juvenile shorebirds are also very approachable; possibly they've just arrived from the Arctic and have had no previous experience with humans. But in most situations, birds actively avoid people, especially people who are moving or making noise.

If a bird you're approaching seems ready to take flight, stop all movement and wait until it visibly relaxes its posture. The closest distance at which you can approach a bird before it takes flight is known as the *flush point.* Slow stalking will get you close to some species, but others are more wary or flighty; that is, their flush points are farther away. With experience you will learn how close you can approach different species for an extended view. It can be hard to predict which flushed bird will move a short distance away and which will fly off into the sunset. If you are birding with a group, it is *very* bad form to overstep the limit and flush the bird that's garnering everyone's attention; the rarer the bird, the bigger the faux pas. Always err on the side of caution in those situations, and know that having a camera does not give you the right to test the flush point without the group members' consensus.

Birds as Birders Birds' sensitivity to intruders, whether predators or humans, ensures their survival. When birds are alerting one another to the presence of a predator, you can heed their warning as well. For example, when small songbirds have discovered a predator like a perched owl or a hawk, they will produce vocal alarm calls and exhibit *mobbing* behavior. Other behaviors that signal the presence of a raptor include a sudden quietness among all the small birds nearby or their sudden dive for cover; pigeons or blackbirds flying in a tight ball; and shorebirds cocking their heads to peer into the sky above them, freezing, or taking flight. Some smaller birds—such as hummingbirds, kingbirds, crows, and even other raptors—have a penchant for harassing flying raptors.

The spacing of birds in a flock on the ground can signal the presence of another species: These birds are aware of the presence of a different species and sometimes avoid it or at least move slightly apart

This **Western Kingbird** *(top)* is mobbing an adult light-morph **Swainson's Hawk**. Even birds as small as hummingbirds will pursue predators that get too close. Most predators rely on stealth and ambush to capture prey, so it's safer for small birds to keep track of a predator's location. Birders can take advantage of these often noisy interactions to help locate birds and observe an interesting behavior. *(Wyoming, June)*

from it. It's always worth carefully checking the edges of a flock, especially if you see an obvious gap between a single bird and the rest of the flock. Even birds that normally flock together can show subtle habitat preferences that segregate them into slightly different groups.

Attracting Birds to You What can you do if you suspect that the movement back in the bushes is made by a bird hidden from view? You can give the bird some time to emerge on its own—this being the least intrusive, and often best, option. If the bird doesn't work its way into view, you can to move to a new position, though you risk flushing it. A third option is to stay where you are and entice the bird into the open with sounds you produce.

■ **Pishing Birds** Birds often respond to intrusion by investigating and then even scolding and mobbing, so birders have learned to attract birds by *pishing*—making rhythmic hissing or shushing sounds that imitate the scolding or alarm calls made by small songbirds when they sense danger. These hissing sounds are hisses broken into a series by opening and closing your lips; shushing sounds are made by pushing out sound between your puckered lips and converting it into a series by the same opening and closing of your lips. Chickadees, titmice, and many sparrows and warblers are likely to react, and if you're near the center of all this attention you can get stunning views. When birds react and make their own scolding sounds, distant and unseen birds are sometimes drawn in as well.

When pishing, start out softly and get louder. A minute or two of pishing will either get a reaction or not; it rarely pays to continue longer without changing your location. Don't overdo it, especially in heavily birded areas or among other birders. Keep in mind that sometimes pishing can have the opposite effect of driving birds away.

■ **Owl Calls** The calls of small owls, the ones that prey on songbirds, elicit a strong response during daylight hours from the very songbirds at risk. Eastern or Western Screech-Owls, Northern Pygmy-Owl, and Northern Saw-whet Owl are all good candidates for imitation. Listen to recordings and seek advice from fellow birders. If you intersperse an occasional owl call with pishing sounds, the songbirds' reaction is sometimes intensified.

■ **Recordings** Many lightweight options for carrying prerecorded bird sounds into the field are available these days. For example, you can buy small MP3 players (iPods and the like) with preloaded recordings and add tiny speakers for playback in the field. Small tape recorders with built-in microphones also allow you to make reasonably good recordings, which you can play back immediately. Guides often use this record-and-playback technique (with a good-quality microphone) on birding tours to entice birds into the open—which may be less intrusive than having the group tromp through habitat for a view—and the bird's response is often immediate. Enticing birds with recordings is intrusive to some degree, more so than pishing or imitating owl calls, because you are entering a bird's territory and creating the illusion that a rival bird is challenging him. So, if recordings are employed at all, use should be minimal.

Describing a Bird's Location Another valuable field skill is communicating to others the location of a bird you're watching. The dilemma is that if you take your binoculars off a bird, you'll probably lose track of it, yet it is almost impossible to describe a bird's location without lowering your binoculars. If you choose to stay focused on the bird, the best instruction you can probably offer is, "Try to look where I'm aiming," and then describe any larger landmarks you might see.

■ **Perched Birds** Begin by describing the location of an object near the bird or the object on which the bird is perched. Note anything distinctive about it, such as, "It's in that close tree with the pale trunk that's leaning to the right," or "It's on the ground near the fifth fence post left of the shed." Give the distance if you can estimate it. Some birders use *binocular field* (the width of your field of view) as a measure of distance to the right or left or of distance above the horizon: for example, "The tree is two binocular fields to the left of the water tower."

Having verbally guided your birding partner onto the right object—
say, a tree—you need to describe where the bird is with descriptions
such as, "Find the first fork to the left, about halfway out from the
trunk." The well-known *clock system* is easy to use, particularly on a
lone tree or bush. Visualize the tree as an upright clock: Halfway up
the tree trunk is the center of the clock, where the hands start; the top
is twelve o'clock, the bottom is six o'clock, and so on. You might say
something like, "It's at ten o'clock, about three feet in from the outside edge, but moving down."

■ **Flying Birds** It is trickier to describe the position of a flying bird.
First, communicate whether the bird is above or below the horizon as
well as the direction in which it's flying. Indicate any distant landmarks—
buildings, distinctive variations in the horizon line, lone trees—and
then add how far above or below the horizon it is. Again, even though
binocular fields vary and are approximate, they are a useful standard,
as in this example: "Look two binocular fields above the white barn,
flying right." You'll need different methods—such as using distinctively

The clock system is used
by birders to describe
the location of a bird
in a tree or shrub.
If you wanted to point
out a specific bird
in this dead tree full
of **Double-crested
Cormorants** you might
say, "Look at two
o'clock, the second bird
in from the right."
(California, January)

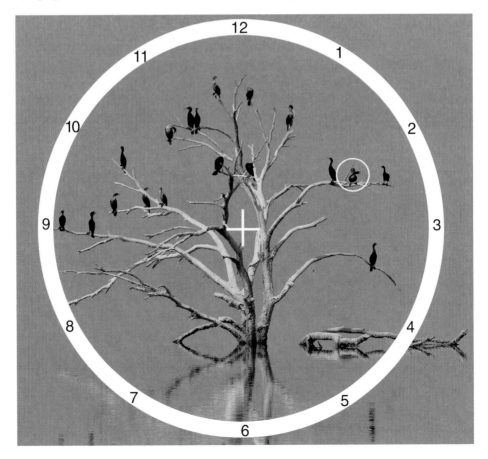

shaped clouds for orientation—to describe high-flying birds. If a bird is high overhead in a blue sky, your best option is to point.

■ **At Sea** On a boat at sea (a pelagic trip) your reference points change, for the boat and the horizon are the only constants. In this case, the boat becomes the clock: The bow of the boat is twelve o'clock, the stern is six o'clock, starboard (right side) is three o'clock, and port (left side) is nine o'clock. The other two essential directions are the distance above or below the horizon and the direction (left or right) in which the bird is flying. You might say, "It's at two o'clock, just above the horizon, flying left."

Birding with Others Employ common sense. Standard etiquette is to keep your voice down and to model the behavior you expect of others. Loud conversations and noisy movements are unacceptable. Choose fabrics that do not make noise when you move, and wear clothes muted in color as well: Minimize contrast by wearing subdued earth tones or darker colors instead of white, which birds find alarming. If you like to attract songbirds with pishing sounds or owl calls, move off by yourself or quietly ask nearby birders whether they object. On a field trip, attracting birds is usually the leader's job, so defer to the leader.

Where and When to Bird

Where to Bird You can bird anywhere and everywhere, but certainly some places are much better than others. Although a myriad of habitat types attract different species, the following general situations apply to most parts of the continent. Later in the chapter, we'll discuss extraordinary North American birding locations and special types of birding, such as owling and pelagic birding.

■ **Edges** Carefully observe habitat edges, which often hold a greater variety of birdlife and activity than areas of uniform habitat. Boundaries between forest and field, hedgerows, watery edges, and coastlines are attractive to many species because they tend to offer a greater variety of food items as well as nearby cover. During migration these edge habitats are particularly favored as feeding and resting areas. In the morning, look for forest edges that catch the first warming rays of the sun. Insects become active earlier in these areas and hungry birds gravitate to them. The same holds true in the evening, so check those areas that catch the last light.

■ **Sheltered Areas** Birds are rarely bothered by light winds, but they seek shelter from strong winds when possible. In open situations, birds shelter by getting on the lee side of any ground irregularity or structure. In vegetated areas the lee side of a windbreak of trees offers shelter and stays warmer. These small, temporary microclimates are worth checking carefully.

■ **Water** Water is a magnet for various birds. The attraction level increases when there is movement in or sound from the water. Any water feature or damp spot, natural or artificial, is worth investigating, especially around its edges. The more arid the surrounding countryside, the more attractive any available water is to residents and migrants alike.

■ **Desert Oases** These habitat islands of the West are bird magnets during migration. Some species are nesters or year-round residents, but oases can be overflowing with migrants for a few weeks in spring and fall. In the 1960s and '70s, Guy McCaskie and others pioneered birding in these areas of California, where they discovered the regular occurrence of eastern vagrants in places like Death Valley and other desert oases. (Prime time for vagrant land birds is after the main passage of regular western migrants.) An analogous *oasis effect* is found on some offshore islands, a welcoming habitat surrounded by salt water.

■ **Sewage Ponds** Sewage ponds can be great places to bird. Be sure both that the facility is open to birders and that it's the right type of facility: You want a two-stage sewage treatment plant, found in many small to midsize communities, which has settling and aeration ponds. (The tertiary plants found in larger cities are less open and less available to birds.) Thanks to the urging of dedicated birders, some enlightened municipalities allow birders to sign in and drive or walk the pond levees. The ponds are incredibly attractive to many species of birds, especially in arid parts of the West where they are often the only bodies of water for miles around. Ducks, shorebirds, gulls, terns, and aerial-feeding swallows are frequent visitors. Many rarities—from Baikal Teal to Gray-tailed Tattler to Ross's Gull—have been found at sewage ponds. Birders who visit these locations may be richly rewarded with the appearance of, for example, Long-tailed Jaeger or Sabine's Gull, which are casual at sewage ponds in the interior West.

This beautiful juvenile **Sabine's Gull** looks like it is swimming in an idyllic setting, but actually the location is a sewage pond in the Mojave Desert of California! Most Sabine's Gulls are highly pelagic during migration, but in fall, they are widespread, though rare, in the interior of North America, frequenting lakes, reservoirs, and sewage ponds. *(California, September)*

■ **Landfills** Although landfills are primarily the domain of gulls and dump trucks, sometimes adventurous birders can be spotted too. As with sewage ponds, contact local officials to see which landfill sites welcome birders. Larger municipal dumps often attract the most birds. When thousands of gulls are present, the odds of finding an unusual gull are greatly increased.

■ **Your Local Patch** Many birders adopt a nearby, prime piece of bird habitat to check on regularly and to cherish. It may be a local pond or park, or a piece of undisturbed marsh or prairie. The study of local birds has lots to recommend it. These species and your intimate knowledge of their occurrence and behavior become your baseline of knowledge when you go farther afield. You can concentrate on specific aspects of their life history, like nesting or feeding behavior, and practice making sketches.

Maintaining your local roots also makes you a valuable advocate for protection and preservation of wild areas. Birders with years of observation records—please keep records!—are often the ones best able to document the importance of an area to wildlife.

■ **Hotlines and the Internet** Birding hotlines and the Internet are great resources for knowing what is going on in your locality. *Hotlines* are prerecorded phone messages—usually updated weekly and hosted by the local chapter of the Audubon Society or the state ornithological society—that detail the when and where of unusual birds, report about what migrants have shown up recently, and give information on field trips and meetings.

The Internet is a vast resource of birding information, but remember that most of it is not reviewed for accuracy. Many sites post local bird news and host forums for sharing information. An important entry portal, *birdingonthe.net,* has links to rare bird alerts, regional servers, and even world news.

When to Bird Dedicated birders, always vigilant, try to make smart decisions about when to bird and how to maximize the chances of seeing more birds or particular species. Timing visits to coastal locations often depends on knowing the local tides; timing a search for migrants is affected by large-scale weather patterns like the passage of a warm or cold front. Local weather conditions can have as great an effect on birding as the changing seasons do.

■ **First Light** Land bird activity usually peaks during the hour or two just after sunrise. Songbirds have high metabolisms and need to feed in the morning. Most migrate at night, which can further deplete their energy stores. Note, though, that adverse weather conditions and low temperatures can curtail early morning activity.

The *dawn chorus* describes the operatic tumult of birdsong that occurs just before and after sunrise during the breeding season. Different species have different starting times, and most species sing a complex *dawn song* that differs from their primary daytime song. Why there is an outpouring of song at this time of day—and why the songs differ—

is unknown. The level of song activity increases again at dusk but not nearly to the level of the dawn chorus.

■ **Tides and Coastal Birds** Many coastal species synchronize their feeding activity with the stage of the tide. Birds feeding on mudflats are particularly attuned to tidal activity. Some of the best shorebirding is done on an incoming tide, when the rising water pushes feeding flocks closer to dry land. If you're positioned correctly, the birds will be pushed closer and closer to you. Low tide is prime feeding time, and birds tend to spread out over a large area; you need a scope in these situations. During high tide look for resting birds and roosting flocks in sheltered areas of dry beach. In complex waterways with numerous bay and inlets, the tidal stages arrive at different times, so you'll need to check a local tide table.

Geographical features sometimes enhance tidal flows or create strong tidal currents. An extreme example of this is the fantastic tidal fluctuation (53 ft.) in the Bay of Fundy in the Northeast. Localized rip currents past capes and around harbor entrances are also enhanced by tidal flow; they tend to concentrate baitfish and can be excellent places to observe flocks of gulls and terns. If you position yourself with a scope on a spit of land that juts out into the ocean or on a coastal headland with enough elevation to look far out to sea, you might be lucky enough to see some truly pelagic species like shearwaters and storm-petrels. Periods of strong onshore winds or major storms and hurricanes can push pelagic species closer to land and increase your chances of seeing them.

This group is birding **Bolivar Flats**, one of the continent's premier shorebird spots, located on the Gulf Coast of Texas. Scopes are a necessity for this type of long-range observation. *(Texas, March)*

Birding Hotspots

The farther you travel from home, the greater the changes in birdlife you'll notice. Distance is just one variable. Habitat and time of year can be equally important. At special places where geography, habitat, and season work together in just the right mix, the result can be extraordinary birding. These notable places—called birding *hotspots*—offer visiting birders the best chances to see regional specialties, avian rarities, or concentrations of migrants.

These special places are not secret. You'll hear about them as you become more aware of the birding possibilities across North America. Excellent site guides exist for many of these locations and are important resources for planning a trip. For example, the *National Geographic Guide to Birding Hotspots of the United States* is an overview of birding hotspots in the continental United States. Many famous locations also have birding festivals during

Miller Canyon in southeast Arizona's Huachuca Mountains offers rare birding opportunities. *(Arizona, August)*

Above-normal high tides occur with regularity: Check the tide tables, which give both tidal elevation and time. Known as *spring tides*—not for the time of year but for the verb meaning "to leap"—they occur near the time of a new or full moon, about every two weeks. This is the best time to observe rails and other secretive marsh-dwelling species that are forced into more open areas by the higher water levels. Storm-related water surges sometimes coincide with spring tides, and this combination can cause exceptional rises in water level, extensive coastal damage, and exceptional birding around coastal marshes.

■ **Weather Fronts and Migration** Weather affects the movements of migratory birds profoundly. Like climate (which is the average pattern

Red-faced Warbler is a specialty of pine/oak canyons in Arizona and southwest New Mexico. *(Arizona, April)*

their best birding seasons. The American Birding Association website *(www.americanbirding.org /festivals)* lists more than 60 festivals, from Florida to Alaska.

Two locations—southeast Arizona and the Rio Grande Valley of Texas—deserve special mention because they harbor so many species that are difficult or impossible to find elsewhere in the United States. Not surprisingly, both of these locations are on the border with Mexico, where the ranges of a number of "Mexican" species extend into the United States.

The mountain ranges south of Tucson are surrounded by desert—earning the nickname "sky islands"—and have strong connections to the birdlife associated with the Sierra Madre in Mexico. The summer nesting season is prime time for visiting, and the hot weather is moderated somewhat by the summer monsoon. In addition to the sky islands' species, many desert-dwelling species can be seen nearby.

In the Rio Grande Valley of Texas, many of the very best known birding spots are located between Falcon Dam and Brownsville. Among the many special birds found in the Valley are outstanding rarities and first North American records. Every November, the Rio Grande Valley Birding Festival in Harlingen attracts hundred of birders (and butterfly watchers as well). ■

Audubon's Oriole is an uncommon resident in south Texas. Look for it near the Falcon Dam. *(Texas, March)*

for a region, not the local weather), the seasonal movement of migrants is predictable and regular—on average, that is—but whether there are any birds to see in your area has more to do with the weather that day and the few previous days. If you want to optimize your birding during migration, pay attention to the weather and learn to interpret it. Catching a major fallout of migratory birds is one of the most exciting birding events on the continent. *(See "Understanding Migration," below.)*

■ **Birding without Binoculars** If you're caught without binoculars, you can still look at birds. In fact, naked-eye birding is great practice, and you'll notice different features—flight style, feeding movements, overall shape and size—that you may have ignored by relying too much on your binoculars. Most small plumage details become irrelevant; only large or high-contrast plumage field marks are usable. When you're out birding with binoculars, spend some of your time looking at birds without them. If you have a cooperative bird that you're ready to leave, make a conscious decision to look at it with unaided eyes for a few minutes. You'll probably learn something new about it. If you do, write it down in your notebook.

Special Types of Birding

■ **Pelagic Birding** Special offshore birding trips are the most common way to see a good selection of pelagic (open-ocean) birds. These include tubenoses—albatrosses, petrels, shearwaters, and storm-petrels—as well as phalaropes, skuas, jaegers, alcids, and pelagic gulls and terns. Some of these species occur in both oceans, but many are restricted to one ocean or the other. To see a wide selection of species, you will have to take trips in a variety of locations and seasons; for rare species, you will need to make numerous trips.

Local birding clubs often arrange pelagic trips, but you'll also find regularly scheduled, commercial trips available in the best locations. The American Birding Association publishes a near-comprehensive list of scheduled trips. The standard trip is a daylong outing and costs about $150. Pelagic trips always have expert leaders onboard. Whale-watching trips and longer ferry trips sometimes offer views of pelagic birds, but they are not as rewarding if your focus is birds.

Masses of seabirds and smooth-as-silk water—pelagic dreams do come true. These happy birders are enjoying the unusual conditions on a **pelagic trip** to Perpetua Bank, about 30 miles off the coast of central Oregon. The distant birds are unidentifiable in this image, but the pelagic species present were mostly Sooty and Pink-footed Shearwaters with good numbers of Black-footed Albatrosses and a Flesh-footed Shearwater. (off Oregon, May)

■ **Owling** While birding during the day, you may occasionally flush a roosting owl. Usually the bird flees before you get a good look at it. A few owl species are active during the day (diurnal), and some nocturnal species occasionally call during daylight hours. To see most species, you will have to make a nocturnal outing, when owls are actively hunting and vocalizing.

Vocalizations are the primary clue to the presence of an owl. Stormy or windy nights make it difficult to hear anything, and most birders think that your chances are better on nights with less moonlight. Since owls are thinly distributed, it's wise to cover as much ground as possible. If you're walking trails, keep your light turned off until you need it and talk only in whispers. When you do encounter an owl, use your light sparingly and don't shine it directly at the bird.

If you hear an owl calling, approach slowly and try to visualize where the owl is before turning on your light. Owl calls can have a ventriloquial quality, so if you're with companions, separate from each other and try to triangulate the location of the sound by pointing in its direction. During the nesting season you should listen for the begging calls of young owls. Sometimes you can see an active owl as it flies across the night sky above you.

Many birders use tape recordings of owl calls to lure owls to them. This is an effective technique, but use it sparingly; owls are easily disturbed. With a little practice, you can learn to produce your own owl calls. Many birders learn to imitate screech-owl calls; these calls are also effective at luring songbirds during the day.

Most species of owls are nocturnal; you'll need a strong light and good ears to locate them. **Whiskered Screech-Owl** is a specialty of southeast Arizona and extreme southwest New Mexico. It is resident in oak and pine/oak canyons between 4,000 and 6,000 feet and is best located by its distinctive calls—a series of short whistles on one pitch or irregular hooting that sounds like Morse code. *(Arizona, May)*

Understanding Migration

Bird migration is among the most awe-inspiring phenomena of the animal kingdom. Twice a year hundreds of millions of birds wing their way either north or south. The migration seasons are certainly the most exciting time to be out birding, and many birders try to organize their schedules around them. Some favored locations are famous migration hotspots: High Island, Texas; Crane Creek, Ohio; and Cape May, New Jersey, to name a few. During the peak of migration season there can be literally thousands of birders in those places. But it isn't necessary to travel to hotspots; migration takes place all across North America. There are bound to be excellent locations to witness migration not far from your own home.

Migration Seasons There are very few weeks during the year when some species or groups of species are *not* migrating somewhere. Migration is the most visible in spring and fall. But consider that some species of shorebirds—for example, adult female Wilson's Phalaropes—have finished breeding and started their southbound migration by the *second week of June,* a time that is, technically, still spring. And the great majority of adult shorebirds have completed their fall migration by the last week of September, before fall has officially begun. On the Great Lakes, gull numbers peak in December, when the winter season starts. Many remain well into January, but as the lakes freeze up, they migrate south or east to open water. Some of these same birds are already on their way back north by the latter half of February. In southern California, northbound Cinnamon Teal and Allen's Hummingbirds have already returned by the middle of January.

Separating **Gray-cheeked** *(top)* from **Bicknell's Thrush** *(bottom)* in the field is not advised except during the breeding season or by vocalizations. The widespread Gray-cheeked breeds across the taiga edge from the Russian Far East to Newfoundland; it winters in South America. Bicknell's has a small breeding range in northeastern North America on mountaintops and coastal headlands; it winters on the Greater Antilles. (top: Connecticut, October; bottom: New Hampshire, June)

Migration is nearly continuous; the timing and the peak of migration depend on the species or group of species involved. In general, waterbird migrations tend to be early in spring and late in fall, whereas passerine migration is later in spring and earlier in fall. For shorebirds, migration spans the entire period, depending on the species involved.

Birds migrate singly and in groups. We've all seen a skein of geese heading north or south; other waterbirds also tend to migrate in flocks. Passerines can form flocks of mixed migrants.

Day or Night Some birds—such as many waterfowl, loons, gulls, and terns—are strictly diurnal (daytime) migrants. Others are largely nocturnal, including most of the passerines, although migration often continues through the early morning hours: Many individuals or small flocks can be seen flying overhead or moving through the trees in the first hour or two after dawn. Some passerines—kingbirds, swallows, corvids—are diurnal migrants. Obviously, diurnal migration is more visible, but nighttime migration can be witnessed by setting up a scope focused on the moon. On a good migration night, you might see the silhouette of a migrating passerine every minute or two, as it passes in front of the moon. The nocturnal flight calls of migratory birds may be audible overhead, particularly during periods of substantial movement. With training and experience, you can identify the flight calls of many species.

Which North American Birds Migrate? As discussed earlier in chapter 3, nearly all bird species show some migratory movement. A few long-distance migrants travel many thousands of miles and do so twice a year; other species only move locally. Very few species, like Wrentit and California Towhee, are almost completely sedentary. Consider also Northern Mockingbird, widely thought to be a resident species. While Northern Mockingbirds are indeed resident within their mapped range, they show some migratory movements, and vagrants have turned up in very distant places—including multiple records from Alaska and the United Kingdom.

An *irruptive species* is one that occasionally moves out of its normal range into areas many hundreds, even thousands of miles away. Most often, these movements occur after the breeding season. Years in which irruptions take place are known as *flight years*. This behavior is usually tied to food shortages within a species' home range; some of these food sources—small rodents for hawks and owls, cone crops for many others—experience regular cycles of abundance and decline, which drive the irruptions. Approximately 30 North American species exhibit some irruptive behavior, and birders are always looking for late summer and early fall movements that might foretell a flight year. Owls, corvids, and finches are the best-represented families, but a few species in other families are also irruptive: ptarmigan, hawks, woodpeckers, chickadees, Red-breasted Nuthatch, Bohemian Waxwing, and Northern Shrike. Populations from one area of the continent may irrupt while other populations of the same species do not. Crossbills and a few other finches may remain for a year or two and breed in the irruptive area.

We think of migration as a behavior of temperate-zone birds that migrate south to spend the colder months in more hospitable areas, often well south of North America. Though not as common, some migratory species in South America regularly move north during their colder months. These are called *austral migrants,* and some—for instance, Fork-tailed and Variegated Flycatchers, and Brown-chested Martin—have occurred in North America as vagrants.

Different Routes for Adults and Juveniles In most species, adults and juveniles may have similar migration routes, even though their timing may be very different. There are exceptions, however. Chapter 3 includes a discussion of the different fall migration routes taken by adult and juvenile Pectoral Sandpipers. With that species and others—like Baird's Sandpiper and American Golden-Plover—juveniles migrate south on a broader front than the adults do. In White-rumped Sandpiper, a casual vagrant in the West (especially unusual in fall), nearly all fall strays have been adults. Sharp-tailed Sandpiper, an Old World species, has dramatically different fall migration routes between adults and juveniles: Adult Sharp-taileds migrate south through the interior of northern Asia; juveniles migrate along the Asian coast. Not surprisingly, the great majority of Sharp-tailed Sandpipers seen in fall in North America are juveniles, including those from Alaska, where hundreds can be found.

Spring Landbird Migration

Eastern Landbird Migration In eastern North America, the routes taken by Neotropical migrants can be described from west to east as circum-Gulf, trans-Gulf, and Florida/West Indies; the spring routes are broadly indicated on the map *(opposite)*. In most cases, one can predict the route taken based on the winter range.

Circum-Gulf migrants travel a land route around the Gulf of Mexico. Many of these migrants winter in eastern Mexico, such as the nominate race of Nashville Warbler. Circum-Gulf migrants can be seen moving in large numbers through south Texas and the Texas Hill Country.

Trans-Gulf migrants fly across the Gulf of Mexico, after stopping and feeding on the Yucatan Peninsula. Most species that winter in Central America or northwest South America—the majority of our eastern Neotropical migrants—are trans-Gulf migrants. The Gulf crossing is perilous, but as long as a "norther" (a cold front with north winds) doesn't arrive during the crossing, most birds make it; and by cutting across the Gulf of Mexico, they avoid many hundreds of miles of travel and other perils. The bulk of the trans-Gulf migration takes place from the upper Texas coast, from Galveston east through coastal Louisiana.

Florida/West Indies migrants take a third route through the Caribbean. Species that winter primarily in the West Indies or eastern South America—for example, Cape May, Black-throated Blue, Prairie, and Blackpoll Warblers—use this route. *Broad-front migrants* are those species with extensive winter ranges that use more than one of the three routes described above. For instance, American Redstart is both a trans-Gulf and a Florida/West Indies migrant; broad-front migrants Black-throated Green Warbler, Black-and-white Warbler, and Northern Waterthrush are numerous on all three routes.

Once they've entered the eastern United States, rested for a bit, and built up their fat reserves, the birds continue north. By knowing the route they traveled into the United States, you can predict the migratory status for many species in your own area. Over the easternmost portions of the Southeast and in the coastal regions of the mid-Atlantic, many species of migrant passerines—except for Florida/West Indies migrants—are rare to almost casual. This holds true until you are north of central New Jersey. The bulk of the spring migration passes to the west of the Appalachians. The spring passerine migration at New Jersey's Cape May has fewer passerine migrants but is great for migrating spring waterbirds and during fall migration. Some Florida/West Indies migrants, like Cape May and Connecticut Warblers, which breed as far west as northeastern British Columbia, start angling west after leaving the Midwest. The volume and variety of passerine migration are most noteworthy in the southern Great Lakes region, where most migration routes converge. Not surprisingly, Crane Creek, Ohio, and Point Pelee, Ontario, are inundated with birds and birders during spring migration.

Western Landbird Migration Western migration is not as well studied as eastern migration. Most western Neotropical migrants don't winter as far south as their eastern counterparts; few venture beyond northern Central America. As shown on the map *(below)*, species breeding in the Rockies and farther east enter the United States in western Texas, New Mexico, and extreme eastern Arizona. Pacific breeding species—Pacific-slope Flycatcher and Hermit Warbler, among others—enter between Arizona and California. Large concentrations of migrating western passerines are reported from the deserts of southeastern California and the coast of northwest Sonora.

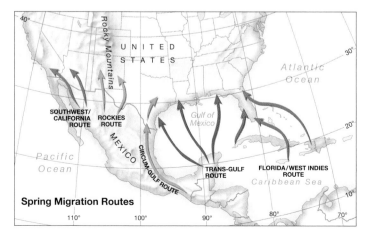

This map shows the main northbound routes that Neotropical migrants take in spring as they enter North America from their southern wintering grounds. For most species, routes shift slightly east during the fall migration, sometimes dramatically so.

This first-fall male **Hooded Warbler** is a classic trans-Gulf migrant. *(New Jersey, October)*

The migration of polytypic species with both eastern and western subspecies is more complicated. For instance, Nashville Warbler has two distinct subspecies with different wintering and breeding grounds. The nominate eastern race *(ruficapilla),* a circum-Gulf migrant, is a common migrant in central Texas, whereas the western subspecies *(ridgwayi),* a Pacific coast states migrant, is a common migrant from southwest Arizona to southern California. In New Mexico, between those two migration corridors, the species is scarce.

Fall Landbird Migration

The most important point about fall migration, in both the East and the West, is this: Migration routes shift eastward in fall. For example, Mourning Warbler is a rare spring migrant in southwest Louisiana and on the upper Texas coast, but it can be fairly common there in fall. Cerulean Warbler and other trans-Gulf migrants are fairly common on the upper Texas coast during the peak of their spring migration, but they are casual there in fall, the migration having shifted eastward. Censuses of migrating birds killed by collisions with radio towers (tower kill data) from Leon County in northwestern Florida report much larger numbers of casualties in fall than spring.

This seasonal shift in migration corridors is also true in the West. For example, Hermit Warblers migrate in spring through the Pacific coastal lowlands and deserts and, in smaller numbers, through the mountains; in fall, their migration shifts eastward and is mainly through the mountains as far east as southwestern New Mexico. The mountain route followed in fall is largely covered in snow in spring and has little to offer in terms of insect food.

Weather and Migration

Weather is a critical determinant in the day-to-day volume of migration. In spring on the Gulf Coast, high pressure and warm weather with winds from the south will result in very few migrants along the immediate coast. But what you see on the ground doesn't necessarily reflect what is occurring thousands of feet overhead. For trans-Gulf migrants, south winds (tailwinds) mean a successful crossing, and most individual birds continue inland for dozens, even hundreds, of miles before stopping near dawn. But if a cold front with cold north winds or wind and rain occurs over the Gulf of Mexico, the crossing becomes a perilous struggle. Rather than being swept up in a warm tailwind, migrating birds must buck headwinds, rain, and cold. If and when they reach the coast, the exhausted birds immediately drop into the first vegetation. This is the classic Gulf Coast *fallout,* an event that birders relish. But in this case, what's good for birders is a disaster for the birds. Many birds fail to reach the coast and perish at sea. In recent years, cold fronts have not been tracking as far south in the latter part of the spring season. It may be a consequence of global warming, but the result is positive: More birds survive the Gulf crossing.

Cold fronts with resulting inclement conditions can also ground large numbers of spring migrants on the Great Plains. Farther north, the passage of a cold front, and especially the aftermath with clear skies and northwest winds, produces very few migrants.

The best conditions for spring migrants in the East seem to be on the back side of a high where the temperature is elevated and the winds are from the southwest. A warm front with warm temperatures, low clouds, and light rain or drizzle is best. These conditions have created some famous fallouts at places like Point Pelee and Crane Creek. These events are memorable, but the best conditions are rare. And unlike the migration spectacles on the Gulf Coast, this weather is not dangerous for the birds.

In fall, cold fronts with northwest winds seem to bring about the best migration conditions throughout eastern North America, especially along the Atlantic coast. A number of species have long offshore, trans-Atlantic flights to South America that are assisted by northwest winds. Among them are certain shorebirds (American Golden-Plover, Hudsonian Godwit, and White-rumped Sandpiper) and some land birds as well (Blackpoll and Connecticut Warblers).

Unique geographical features in certain locations alter the general rules of what is the best weather for seeing birds. Ontario's Point Pelee extends into Lake Erie from the north and funnels fall migrants to its tip. A wind from the south will stop migrants from crossing the water into Ohio and concentrate them at the point. As a rule, however, the best fall migration birding in the East is after a cold front.

In general, migration in the West appears to proceed at a steadier flow and with fewer pronounced concentrations of birds than in the East. Along the Pacific coast an inversion layer often persists from mid-spring to early fall. Landbirds migrating along the coast fly above the inversion and then drop below it at dawn. Some find themselves over the Pacific Ocean, so they then backtrack to shore. These weather conditions offer birders a chance to see the largest number of migrants at coastal locations. Another condition in the West that can produce large numbers of migrants happens in fall. Strong high pressure builds up over the Great Basin, and sometimes it results in downslope, north-easterly Santa Ana winds. These strong winds, responsible for fanning California's famous brushfires, bring migrants from the interior. Often the best birding occurs when the winds die down a bit and some coastal fog appears.

Migration Specialties

We have detailed general birding opportunities during migration, but various specialty niches of migration watching deserve their own discussion. Some of these activities, such as hawkwatching, have become very popular activities, attracting thousands of dedicated observers.

Hawkwatching Hawks and falcons are diurnal migrating raptors. With the exception of some falcons (Peregrine Falcon and Merlin), raptors have a strong aversion to crossing water and will fly well out of the way to avoid such crossings. Certain coastal and Great Lakes promontories are well-known raptor migration sites where concentrations of birds regularly occur. Points on the south sides of the Great Lakes are best in spring; spots on the north side of Lake Erie are best in fall. Some migrating raptors, like Broad-winged Hawk in the East and Swainson's Hawk in the West, can be encountered in the hundreds, even thousands. In addition to visiting these headlands, raptors also use mountain ridges during migration because they provide favorable *updrafts* when a prevailing wind collides with a ridge and is deflected upward. Updrafts and *thermals* (local updrafts of ascending hot air caused by solar heating of the ground, which rarely form before midmorning) carry circling raptors effortlessly upward. Once a bird has soared to a high elevation, it can leave the thermal and glide in any direction. As it glides, it slowly loses altitude, but this glide can take it many miles; when the next thermal occurs, the sequence starts over.

Seawatching and Waterbird Migration Because many of the locations described below require that birders view the birds from a great distance, it is best to use a scope in addition to binoculars *(see page 203)*. Without a scope, you will not be able to identify some distant birds.

■ **Coastal Headlands** A number of popular sites along both the Atlantic and Pacific coasts are ideal for watching the migration of various groups of waterfowl, loons, gulls, and sometimes shearwaters, jaegers, and alcids. A strong onshore wind is ideal for pushing more pelagic species closer to shore. Many waterbirds are coastal migrants but stay offshore rather than follow the ins and outs of the coastline and only come close to land when they encounter a headland jutting into their path.

■ **Waterbirds on the Great Lakes** Large numbers of waterbirds, including waterfowl and gulls, are found on the Great Lakes. Fieldwork in recent decades has shown that the most dramatic lake watching takes place in fall at the south end of north-south lakes. Unusual species, including jaegers and rare gulls like Sabine's Gull and Black-legged Kittiwake, are regularly noted in fall. Birding at these sites, especially with strong, northwest winds after the passage of a cold front, offers some of the year's most exciting birding.

■ **Inland Reservoirs** Inland lakes and reservoirs, commonplace in North America, are important sites to see waterbirds. The ongoing passage of birds may not be as obvious as it is at a coastal headland or at the best spots on the Great Lakes, but you can appreciate the volume of migration by noting the day-to-day changes. Inclement weather offers the best conditions to see large numbers of waterbirds. Check your local reservoir during and immediately after the passage of a front to see the largest number and variety of birds.

Ross's Gull is an exciting find no matter where you are. These petite gulls are generally found in high northern latitudes, but there is a scattering of records for southern Canada and the lower 48. This adult at the south end of the Salton Sea in 2006 established the world's most southerly record for a Ross's and was the first record for California. *(California, November)*

■ **Tropical Storms** In recent years, storm chasing has become a specialized birding pursuit. Not unlike tornado chasers, these birders chase hurricanes in hopes of seeing unusual pelagic species on the coast or even inland. Species such as Black-capped Petrel have occurred as far inland as Lake Erie after the passage of a hurricane. Checking the coast during or after a hurricane is often not as productive as checking inland reservoirs or points on the Great Lakes, for two reasons. First, reservoirs, lakes, and rivers concentrate displaced birds into an area much smaller than miles and miles of coastline. Second, being on the coast can be dangerous during and after a major storm, and many areas will be closed to travel or evacuated. Major storms rapidly lose their force as they move onshore, so inland areas are usually easier and safer to access. The general consensus among storm-birders is that the most productive location for finding storm-driven birds is on the east side of the eye of a tropical storm.

Birds driven inland by storms can reorient and attempt to return to sea. The artificial islands that are part of the Chesapeake Bay Bridge and Tunnel in Virginia are a favorite place to witness this phenomenon. After a hurricane, you can observe pelagic birds as they pass by, exiting the Chesapeake Bay and returning to the Atlantic Ocean. Recent hurricanes have produced records of petrels, shearwaters, storm-petrels, tropicbirds, frigatebirds, and tropical terns at this location.

Rare Birds

Finding a rarity is always exciting. It also challenges your overall birding skills. Below are some tips to help you find unusual species.

Understanding Vagrancy Patterns As with common birds, most rare birds demonstrate patterns of occurrence. Very few records stand alone. Learning patterns of occurrence in advance gives you an idea of where and when to look for rare birds. Keep in mind that a resident bird out of range—for example, a normally sedentary Rufous-crowned Sparrow turning up at a desert oasis—may be much more unusual than a highly migratory, casual vagrant.

A rarity often produces a mass "twitch." These observers are lined up to see Rhode Island's first record of Long-billed Murrelet in November 1998. This species breeds in coastal northeastern Asia, but records are scattered across North America and there are even two records for Western Europe.

Looking for the Unusual Rarities are typically found by looking through common species for anything unusual. For instance, while scanning through a flock of Bonaparte's Gulls, look for a similar but slightly larger bird with dark undersides to the primaries. It might be a Black-headed Gull, a scarce Eurasian species. If you find a gull smaller than a Bonaparte's that has dark underwings, you likely have

an adult Little Gull; if it is smaller and also has a bold black carpal bar, it's likely an immature Little Gull. When scanning through a flock of scaup, look for any that might have a black back. It may be a Ring-necked Duck, but it could also be a rare Tufted Duck. If you encounter a flock of Chipping Sparrows, be on the lookout for a bird of comparable size and shape that is buffier overall with pale lores and a more striking head pattern. It could be a Clay-colored Sparrow, a rarity on both the East and West coasts. Again, if you know your local birds well, a rare bird will stand out from them in some way.

Being Cautious When you find a rarity, don't celebrate immediately. First, ask yourself whether you could be wrong. Recheck all of the field marks carefully and then check them again. A rarity should be seen well and seen, preferably, *over a prolonged period of time.* A briefly seen "rarity" usually leads to a troublesome identification, so be slow to claim a rare bird after an overly brief sighting. A sure sign of trouble is when you can't find again the rare species you think you just saw. Say you are looking through a flock of peeps and suddenly think you've spotted a Little Stint. But just then the flock flies off, circles, and lands again, and now you can't find the Little Stint. Even when one or two birds flew off, you should reconsider your identification. *(See page 39 to learn more about what to do when you find a rare bird.)*

Finding a rarity often requires looking through masses of a more common species. Here, an adult **Black-headed Gull** (center) is flying among a flock of adult **Bonaparte's Gulls**. Note Black-headed's larger size and the blackish shading to the underside of the primaries. European observers would be faced with the reverse situation—looking through many Black-headed Gulls for a Bonaparte's. (New Jersey, February)

This **Western Reef-Heron**, standing in front of a **Snowy Egret**, is one of a very few to ever to appear in North America. It established the first record for both Maine and New Hampshire. It is believed to be the same individual that was present earlier in the summer of 2006 in Nova Scotia and perhaps the previous summer in Newfoundland. *(August, New Hampshire)*

Chasing Rare Birds Finding out about rare sightings is much easier with the advent of hotlines and timely Internet reports. Interested birders can descend on the location of a rare sighting in a very short time. Recently, a Western Reef-Heron *(left)* in New Hampshire and southern Maine—one of very few ever recorded in North America—was seen by thousands of birders across the country.

A chase for a major rarity often resembles a social event more than a birding expedition. It can be enjoyable, but don't let your observations become less discerning. It's best to approach the task as though everyone else may be wrong; look at the bird as if you were the first one to discover it. There are countless incidents of a rarity seen by dozens, even hundreds of birders, that is eventually proven to be a more common species. It is important always to make your own critical evaluation.

Keeping Track

Listing Birders are list makers. You may keep a North American list, a state or provincial list, or a county list. The size of one's list indicates only the amount of effort and time spent pursuing birds in an area, not your skill as a birder. Many birders use computer software programs to keep track of their lists, whereas others prefer to mark a checklist or to keep a card file.

Keeping a Journal Much more important than compiling lists is keeping a journal to record what you see. Journal entries should include both a list of species seen and an estimate of the numbers of individuals you saw. It should also include the specific circumstances of your day's outing (date, weather, time spent, names of those birding with you), and each location should be a separate entry. In addition to describing any unusual or rare birds you see, you might choose to include notes on common birds: vocalizations, behavior, and aspects of their appearance.

Taking Photographs Bird photography has always been a popular avocation and now, with digital camera technology, it is both easier and less expensive. Digital photography through a birding scope, known as digiscoping *(see page 206),* has become an increasingly popular and effective way of documenting rarities. Your collection of photographs will also supplement your field notes.

eBird The Cornell Laboratory of Ornithology and the National Audubon Society launched eBird in 2002 at *ebird.org* to allow birders throughout North America to share their observations via the Internet in a massive compendium of bird sightings. Observers enter data from their checklists for specific birding locations. The resulting real-time database tracks the distribution, seasonal status, and long-term population trends of all North American species. The program can generate range maps and seasonal graphs of species occurrences; it also stores your personal birding records for your own use. In 2006, eBird participants reported more than 4.3 million bird observations across North America.

Volunteering Many volunteer opportunities offer a chance to help the birding and environmental communities and to increase your birding knowledge. Volunteering to lead field trips for your local chapter of the Audubon Society or another birding organization is one such activity. Leading trips puts you in the position of being a teacher and hones your skills. You don't have to be an expert to lead trips—intermediate-level birders are often the best teachers for beginners. When you become a veteran birder, consider mentoring one or more individuals.

Several notable programs, listed below, use large numbers of volunteer birders in the role of *citizen scientists*. This large and growing pool of observers gathers huge amounts of data—a task that simply could never be accomplished by the relatively few professional ornithologists and wildlife managers.

■ **Christmas Bird Counts** These counts have a long, storied history in the birding world. The first one, held in 1900, had 27 participants;

Bird observatories are a major resource for research and provide opportunities for volunteers. Here banders at the biological field station of the Carnegie Museum of Natural History at Powdermill Nature Reserve in southwest Pennsylvania are studying the pattern of the primary coverts on Blue Jays in order to properly age them.

Field Sketching

Making sketches in the field is a good way to focus your observations and it's one of the fastest ways to learn about birds. The act of drawing a bird requires that you look closely at all aspects of its plumage and structure. Your sketches can also be used to document any rare or unusual sightings. Don't be discouraged by thinking that you lack artistic ability: For most birders the sketching process is not about being an artist, it's about being an observer. In addition, most birders find that their sketches improve dramatically after some initial attempts.

Start out with some stationary models if you can; waterfowl, egrets, or roosting gulls are large and cooperative subjects. A birding scope allows you to stay focused on a subject and keeps your hands free. If your scope has an angled eyepiece, you can position your sketchbook in a way that lets you easily alternate between the view of the bird and your drawing without moving your head. Simply switch between your left eye and right eye. In the field, I use a mechanical pencil with a built-in white (smudge-free) eraser and a 5-by-8-inch stitched notebook.

The following is a simplified, four-step process to get you started. I've sketched a songbird—a Carolina Chickadee—as an example. The sketches show the progression of a single drawing. A side view is often the easiest pose to start with, and it gets the most information into a single drawing. To get the feel of the process, take a piece of paper and follow along by duplicating the sketches below. Nonpasserine birds have more varied shapes, but the same building blocks and techniques work equally well for them.

1. Draw the Basic Shapes Start with two ovals—a large one for the body and a smaller one for the head. Concentrate on getting the proper size relationship between the two ovals. Before you draw the head oval, think about the length of the neck. Most songbirds have short necks that are hidden when the bird is relaxed, so the head oval overlaps the body oval. This is the case with chickadees, which also have large heads in relation to their bodies. Long-necked birds, like egrets or cormorants, would have the head connected to the body with a tubular neck. At the rear end, a wedge

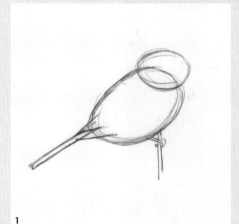

1

2

shape—for the uppertail and undertail coverts—connects the planklike tail to the body. Check the proportions of these basic shapes as you proceed. Keep it loose and sketchy, and don't hesitate to erase if the balance of shapes seems off.

2. Refine the Shapes Start to refine the outside shape with transitional lines at the throat and the back of the head, and make your first indication of the bill and eye. In most species the eyes fall close to the bill, not in the center of the head. Indicate where the legs come out of the body; keep it simple, though, and don't get too involved with the feet. Add another basic oval shape for the folded wing. Erase any extra lines.

3. Add Details Keep refining your shapes and start to add feather details. A basic understanding of bird topography is very helpful at this stage. If you need to refresh your knowledge of the parts of a bird, study the annotated photographs in chapter 4. Particularly important is the relationship of the various feather groups of the wing—primaries, secondaries, tertials,

coverts, and scapulars (covering the upper part of the wing). Look at details, such as how the greater covert feathers overlap in one direction and the median covert feathers (directly above them) overlap in the opposite direction. The primary tips usually project beyond the basic oval shape of the wing.

4. Add Patterns and Annotations After the feather outlines and the structure of your bird are in place, you can add any patterns by shading. To add detail to your drawing, look at specific areas of the live bird, memorize what you see, and then transfer those details to your drawing. A chickadee's pattern is simple and bold, giving the bird its distinctive appearance. Write annotations on color and point out salient features you noticed. If you are documenting a rare bird, you'll probably want to write extensive notes. You can add watercolor washes or colored pencil shading to your sketches when you get home. You learn about the bird in the drawing process, and the sketches in your journals will also become a valuable personal reference. ■

— *Jonathan Alderfer*

3

4

The **Christmas Bird Count** has been a popular pastime for over a century and contributes much useful data on the early winter distribution of North American birds. These birders posed for the camera during the Tulsa, Oklahoma, Christmas Bird Count in December 1936.

there are now almost 2,000 separate counts, with more than 50,000 participants. The scope and longevity of these surveys, which also are great fun, make them valuable contributions to the citizen-scientist monitoring of winter bird populations.

■ **Breeding Bird Survey (BBS)** Every year volunteers run thousands of survey routes across North America to compile a census of breeding birds. These established routes are run at the height of the breeding season (June, in most locations). Each route is 24.5 miles long, with survey stops every half-mile. At each stop, volunteers record every bird seen or heard in a three-minute window. Started in 1966, the BBS is coordinated by the Patuxent Wildlife Research Center of the U.S. Geological Survey and the Canadian Wildlife Service. Volunteers need to have good hearing and be able to recognize the bird sounds they are likely to encounter on their route. The data is used by organizations and publications that monitor the health and population trends of North American birds. More information is available online at *www.pwrc.usgs.gov/BBS*.

■ **Atlasing** A *breeding bird atlas* is a snapshot of the status and distribution of breeding birds in a certain area, such as a state or province, which is divided into many survey blocks using a grid. (There are also nonbreeding bird atlases.) Compiling this data takes the combined efforts of hundreds of volunteers at all skill levels. Most atlas projects entail gathering data over a period of years (usually five), after which the work is published online or as a book.

Your Birding Library

Every committed birder amasses a personal birding library. You can begin building your collection by acquiring field guides and status and distribution references to the birds of your area. Develop a system for organizing your references, for it is important to be able to find a particular reference quickly.

Magazines and Journals Membership in your local chapter of the Audubon Society often includes a subscription to the chapter's newsletter—a great source for local birding news. You should also subscribe to your state or provincial journal and keep it for reference. Several national magazines and journals are also well worth your consideration.

■ **North American Birds** The American Birding Association's *North American Birds* is an indispensable quarterly journal of seasonal status and distribution that covers all of North and Middle America. The main body of the journal, organized into 34 regional reports, attempts to mention all significant records within the season covered. Articles on first records, distribution, and identification—as well as pictorial highlights—are also published in every issue.

■ **Birding** The flagship magazine of the American Birding Association, *Birding* features timely birding news, identification articles, and a lively mix of product and book reviews.

■ **Living Bird** *Living Bird,* the quarterly journal of the renowned Cornell Lab of Ornithology, focuses on bird conservation issues and reporting on citizen science projects.

■ **Western Birds** The quarterly journal *Western Birds,* published by Western Field Ornithologists, includes excellent peer-reviewed studies on the bird distribution, ecology, and identification that interest those living in western North America.

■ **Other Popular Magazines** There are less-technical birding magazines that many birders enjoy. These include *Bird Watcher's Digest, Wild Bird,* and *Birder's World.* They offer articles on a myriad of popular subjects, such as birdfeeding, identification, travel destinations, and product reviews. Human-interest stories are also featured.

■ **Ornithological Journals** In North America, five well-known ornithological societies publish quarterly journals containing peer-reviewed, scientific articles. Beginning birders probably won't want to join these organizations, but they may encounter references to articles published in these journals. Past issues are archived online by Searchable Ornithological Research Archive (SORA) and available for free at *elibrary.unm.edu/sora/index.php.* To maintain their subscriber bases, the organizations do not make the most recent years' issues available.

    The Auk is a publication of the American Ornithologists' Union. The annual supplement to the *Check-list of North American Birds* appears in this journal and is the standard reference for all changes in taxonomy and nomenclature. *The Condor* is a publication of the Cooper Ornithological Society; the *Wilson Bulletin* is a publication of the Wilson Ornithological Society; *Journal of Field Ornithology* is a publication of the Association of Field Ornithologists; and *Waterbirds* is a publication of the Waterbird Society.

■ **European Publications** Many birding magazines and journals from Europe are published in English; outstanding examples include *British Birds, Birding World, Dutch Birding,* and *Alula.* There is substantial overlap in birds that occur in the Nearctic and Palearctic regions, especially among the nonpasserine species, and European birders have been on the forefront of working out challenging identifications in such groups as waterfowl, loons, shorebirds, gulls, and jaegers. The majority of their efforts are published in the journals listed above: If you have the interest, a subscription to any of them will be well worth your investment.

Books Books on birding number in the thousands. We cover the main categories and make a few recommendations, but many excellent choices go unmentioned by specific title due to space constraints. *(See also "Additional Reading," page 218.)*

■ **General Reference** Every birder should have a one-volume, general reference to birds with easy-to-use, alphabetized entries. Two excellent ones are *A Dictionary of Birds* and *The Birdwatcher's Companion to North American Birdlife.* The American Ornithologists' Union and the American Birding Association publish annotated checklists covering all of North America. For those who wish to delve into the intricacies of North American subspecies, the bird-banding handbook *Identification Guide to North American Birds,* part 1, is the essential reference.

For world travelers, a complete checklist of all the world's birds is essential, but checklists must be updated when taxonomic changes are made. (Listing software is an alternative.) The magnificent series *Handbook of the Birds of the World* has published 11 out of a projected 16 volumes. Every species in the world will be illustrated, and the text for every family is comprehensive and includes many stunning photographs—an incredible, though expensive, resource.

On a North American scale, *The Birds of North America* series covers all the *breeding* birds north of Mexico, including Hawaii. Originally published as separate pamphlets for each species, it is now available online (with video and sound recording) for a modest yearly fee at *bna.birds.cornell.edu/BNA.*

■ **Field Guides** The two authors of this book are the editors of *National Geographic's Field Guide to the Birds of North America,* fifth edition. Most birders acquire more than one field guide, and numerous choices are available. See the section in chapter 2 about field guides *(page 25)* for a discussion of features to look for. A European field guide adds a different and useful perspective; *The Birds of Europe* is an excellent choice.

■ **Status and Distribution** This group of books is discussed in chapter 3 *(see page 37)*. In addition to state and regional books, many locations, such as national parks and refuges, have checklists. Those that include bar graphs are the most useful. At a minimum, you should purchase a status and distribution guide for your state or province.

■ **Guides to Families** Specific groups or families of birds have field guides devoted to their identification. Many of these works have a worldwide scope. The pioneering books (from British authors) in this genre are *Seabirds* and *Shorebirds*. Many other excellent examples have followed. There may be a guide to a family that particularly interests you; some birders collect all the books that specifically cover a family of interest. Other family guides have a regional, rather than worldwide, scope.

Some of the guides for families in North America include *A Field Guide to Hummingbirds of North America* and *Hummingbirds of North America*; *A Field Guide to Warblers of North America*; and *Sparrows of the United States and Canada*. Family guides for North American hawks and shorebirds are too numerous to list here.

■ **Site Guides** Because birders are inveterate travelers, site guides exist for every state and province and for many counties as well. Most site guides are very specific, giving driving routes and mileages, lists of possible species, and seasonal occurrence. The American Birding Association publishes an excellent series of site guides that cover many of the premier birding locations in North America. There's even a recent one to North American metropolitan areas.

■ **Foreign Field Guides** World travelers need not despair. These days there are field guides to most of the regions of the world published in English, and new titles are published every year.

■ **Sound Recordings** Most birders acquire a set of CDs that features the songs and calls of North American birds; these are usually available as both East and West compilations. An excellent recent book, *The Singing Life of Birds,* accompanied by its own CD, introduces the reader to the science of birdsong.

Additional Equipment

One of the charms of birding is how unfettered and natural the experience can be. Equipment is minimal—binoculars are the only real necessity—but there are add-ons that many birders find essential. A scope (birding telescope or spotting scope) and tripod are truly essential for some types of birding. The rest of the gear mentioned below is helpful and fun to use but of secondary importance.

Birding Scopes A birding scope is a relatively compact, single-barrel optical instrument able to produce far greater magnification than binoculars: 20x–60x magnifications are common. Scopes are not substitutes for binoculars; they are used in addition to them. Using a scope requires a tripod to hold it steady and a birder willing to carry it into the field. Because a scope and tripod are heavy (together they can weigh around ten pounds) and somewhat clumsy to carry and set up, it takes some commitment to use them. The payoff comes when you get stunning, close-up views of distant birds that are little more than specks in binoculars. Scopes are at their most useful when you're looking at distant birds in open situations—shorebirds on a mudflat, ducks on a lake, gulls on a distant levee; they are difficult, but not impossible, to use on small landbirds. Some birders cannot be bothered with the hassle of carrying and setting up a scope, whereas others cannot bird without one.

■ **Choosing a Scope** Much of the advice in chapter 2 on choosing binoculars applies to choosing a scope. Unless you already know exactly what you want, try different scopes before you choose one, and buy from a reputable retailer. Get all the advice and recommendations you can from fellow birders. Most will be happy to share a view through their scopes and regale you with details about what they like and dislike about them. This is valuable information, but you have to balance it with solid figures on price and specifications. Read reviews in popular magazines or online; most of them will have comparative charts featuring the specifications of many makes and models. Prices range from $300 to well over $2,000. As with binoculars, you get what you pay for.

■ **Size** The size of a scope is a function of the size of the objective (front) lens. Full-size scopes usually have 80-mm objective lenses, compact scopes' lenses are about 60 mm. A larger objective lens will offer up a brighter image, but the corresponding scope will be heavier. Very high-end (expensive) compact scopes are vast improvements over their predecessors, and many birders have made or are considering making the switch to one because compacts are much lighter and easier to carry. The best scope in the world is useless if you can't be bothered to carry it with you.

Two popular scope designs, a **straight-through scope** (left) and an **angled scope** (right) are shown. Angled scopes can be particularly good for tracking flying birds, and their lower position makes them easier to share with viewers of different heights.

Digiscoping has become increasingly popular in recent years. It can be done using attachments or by simply holding a small digital camera directly up to the scope's eyepiece, as shown here. It is an important tool for documenting rare birds.

■ **Power** The magnification or power of your scope depends on the eyepiece with which it is fitted. Most manufacturers offer a choice of eyepieces, which are sold separately from the scope body. Zoom or fixed-power eyepieces of various magnifications are available. Most birders choose a zoom lens for its versatility—20x through 60x is standard—but the field of view is rarely as expansive as a wide-angle, fixed-power eyepiece. Many experienced birders prefer a 30x wide-angle eyepiece as their primary lens. The wide field of view allows for easy scanning and a restful image.

■ **Angled or Straight-through** This refers to the arrangement of the eyepiece *(opposite),* and there are advantages to both designs. Angled scopes are becoming popular for several reasons: They allow viewers of differing heights to share a single scope, they are good for long-term viewing, they allow you to look over the eyepiece, and they are easier to use on soaring birds. You can usually get by with a short, lightweight tripod. Straight-through eyepieces are the popular favorites: They are easier to aim, the eyepiece is easier to shield from rain, they have a higher viewing height (allowing you to look over low bushes or fences), and you can use them with car-window mounts. Try both designs before you decide. Most manufacturers offer both.

■ **Tripods** You have to mount your scope on a tripod, and this is not a good place to try to save money. Expect to spend $150–$300. A rugged and stable tripod is essential to get the most out of your scope because movement or vibration kills a good image. Some tripods are specially designed for birding use and have different features than photography tripods, although there is overlap. The lightest tripods that maintain good stability are made from carbon fiber, but they are expensive. Aluminum tripods are the standard. Look for flip locks on the legs extensions for quick setup and an anodized black finish, which is better camouflage than shiny aluminum. A fluid head (for very smooth panning) with a quick-release plate is a good choice for attaching the scope to the tripod.

■ **Digiscoping** The term *digiscoping* refers to using a compact digital camera to take pictures through a birding scope. The practice has become popular in the birding community over the last decade, and it offers a low-cost method for documenting your birding discoveries. Many birders are casual digiscopers: They carry a small digital camera in case they run into something interesting or need to document a rarity.

This hybrid male **Mountain x Western Bluebird** was digiscoped with no attachments in Round Valley, California, near Bishop, on December 17, 2006. The image documents the appearance of this almost unknown hybrid combination. The bird resembles a male Mountain Bluebird, except for warm brown shading on the back and sides. The center of the breast, although not visible here, is orangish. *(California, December)*

At the most basic level, all you do is hold your camera's lens up to the eyepiece of your scope and press the shutter release. In a pinch, you can even hold a camera up to your binoculars. Handheld results can be quite good. Use the lowest magnification on your scope, and zoom the lens on your camera a little to help reduce vignetting. Practice holding and aligning the camera, and take lots of images. Experiment to your heart's content; the images are free once you own the equipment.

Many digiscopers take the next step and buy an adapter that holds the camera in perfect alignment with the scope, reducing camera shake. Images are vastly improved by using an adapter. Most adapters require threads on the camera lens for attachment. The downside is that adapters and other associated gear can get in the way when you are "just birding." If you wish to explore digiscoping at a more advanced level, do some online research. The following websites are helpful starting places: *www.digibird.com; www.digiscoping.ca;* and *birdingonthe.net /mailinglists/DIGI.html.*

Other Gear Chapter 5 includes a short discussion on using tape and digital sound recorders while birding, with the various options of buying prerecorded birdsongs or of recording in the field *(see page 123).* Like photography, this is a specialized subject that you can pursue on your own. Some other gadgets with a practical use in the field or at home are mentioned below.

■ **Personal Digital Assistant (PDA)** Software has been developed recently that turns a PDA, such as a Palm Pilot, into an electronic field guide. *National Geographic Handheld Birds,* a portable interactive field guide, displays the text, art, and range maps of its well-known field guide on a PDA's full-color LCD screen. Its scroll and search functions make it easy to use, and the audio function has recordings of songs and calls that you can listen to in the field. An online demonstration is at *handheldbirds.com.*

■ **Global Positioning System (GPS)** Most birders know about the small pocket-size GPS devices that allow you to document your exact position by longitude and latitude using satellite technology. This is handy for guiding a fellow birder (if he also has a GPS) to the exact spot where you found something of interest. It also allows you to precisely document your sighting locations for submission to eBird, although submissions do not require such precision.

■ **DeLorme Atlases** These handy paperback atlases, available for most states, are based on topographic maps. They show roads, elevations, rivers, lakes, and even sewage ponds. Most birders carry a copy of the atlas for their home state in the car, and some birding hotlines and Internet sites include references to these maps when giving directions.

■ **Listing Software** Although we have no personal experience with these products, many birders find them very useful for keeping track of their various lists. The most popular programs are *AviSys* (Perceptive Systems), *BirdBase* (Santa Barbara Software), *Birdbrain* (Ideaform), and *Birder's Diary* (Jones Technologies). Expect to pay about $100 for each program. You can research the various features online, but note that of the programs listed, only *Birdbrain* works on Apple computers.

Always a crowd pleaser, **Great Gray Owls** are popular subjects to photograph. This boreal species moves into more southern and eastern locations during some winters, probably due to food shortages. *(Ontario, March)*

CHAPTER 9

TAXONOMY AND NOMENCLATURE

B iological classifications bring order to the complex diversity of life on the planet. Standardized taxonomy attempts to express the evolutionary relationships of organisms and to provide them with a stable set of names. Living organisms are classified in a hierarchy of taxonomic units which, in decreasing inclusiveness, are *kingdom, phylum, class, order, family, genus, species,* and *subspecies*. Other designations (such as *subfamily* and *tribe* for levels below family but above genus) are often recognized and used by taxonomists. Within the animal kingdom, birds are in the phylum Chordata (and the subphylum Vertebrata, the backboned animals) and in the class Aves. Within North America, north of Mexico, more than 80 families of birds in 22 orders have been recorded. The largest avian order is certainly the Passeriformes, our perching birds, which includes all birds known as songbirds. The taxonomic authority for North American birds is the American Ornithologists' Union's Committee on Classification and Nomenclature, and its most recently published taxonomic checklist is the seventh edition of the *Check-List of North American Birds* (1998). Supplements to that work are now published annually in the July issue of the AOU's quarterly journal, the *Auk*.

The Swedish botanist **Carolus Linnaeus** *(above)* is the father of our binomial system of scientific names, a system he devised in the mid-18th century. **Sooty Grouse** *(opposite)*, found from southeast Alaska to central California, was split from the more interior Dusky Grouse in 2006. Both species were formerly known as the Blue Grouse. *(California, November)*

The Linnaean System

We owe our current system of scientific names to Carolus Linnaeus. Prior to the mid-18th century, there were many different names that could be applied to the same species of organism. This reflected that names were applied at different times and often by those speaking different languages. Linnaeus wished to have a uniform scientific language that could be used by all, no matter which language they spoke. He gave all named species a *scientific name* composed of two parts: a *genus* (always capitalized) for the first part and a *specific epithet* (not

capitalized) for the second part. Thus Magnolia Warbler is known by the scientific name of *Dendroica magnolia*. Though often referred to as Latin names, scientific names are derived not only from Latin, but also from Greek. All plants named prior to 1753 and all animals, including birds, named prior to 1758—including those that Linnaeus had named himself in his work *Systema Naturae,* first published in 1735—were renamed in a system that we still follow.

Subspecies Nearly a century after Linnaeus published his epic work, the concept of naming subspecies within a species came about. This resulted in the necessity of using trinomials for polytypic species (variable species that have multiple named subspecies). While the idea originated in Europe, American ornithologists of the late 19th century were the ones who embraced it and applied it to their own publications where new species and subspecies were being named. This is discussed extensively in chapter 6. Briefly, if the species is monotypic with no named subspecies (*race* and *form* are synonyms), it is only known by the binomial. Thus Magnolia Warbler (a monotypic species with no recognized subspecies) is *Dendroica magnolia*. A species does not become polytypic until a new variety (a subspecies) of that species is described. As we saw in chapter 6 with Palm Warbler, the first variety of the species named becomes the nominate race. This results in the second and third parts of the scientific name being identical. Thus, the more westerly variety of Palm Warbler (the first named) is *Dendroica palmarum palmarum*; the next named subspecies became *Dendroica palmarum hypochrysea.* (The subspecies name, like the specific epithet, is not capitalized.) New species and subspecies described are almost always backed by specimens, and one such is always designated as the *type specimen.* Type specimens are very valuable for taxonomists and are carefully preserved so that others, if needed, may re-examine them at a later date.

Investigating which subspecies makes up the nominate race teaches you something about the species' ornithological history. Species that are found across the Northern Hemisphere are likely to have a nominate race that comes from Europe, even Scandinavia. In North America, the eastern race is usually the nominate, reflecting where most of the earlier ornithological work was done. There are exceptions, though: For example, the nominate race of Hermit Thrush is a western subspecies, *Catharus guttatus guttatus,* reflecting Russian research done in the early 19th century in Alaska. Surprisingly, the widespread Red-tailed Hawk was described to science in 1788 by Johann Gmelin from birds on Jamaica. The nominate race, *jamaicensis,* is restricted to that island.

Again, as we have seen in chapter 6, exactly which populations to recognize formally as subspecies and how to draw boundaries between subspecies are subjects of ongoing debate, as is the concept of subspecies itself. However, geographical variation within a species is a matter of great importance—both because of sometimes significantly different appearances and because of conservation issues associated with evolutionary diversity. Recognizing and protecting this diversity is surely a worthwhile goal. As we have lost some North American species over the last 160 years (e.g., Great Auk, Labrador Duck, Passenger Pigeon, and Carolina Parakeet), we have also lost various subspecies. One such subspecies was "Heath Hen" *(Tympanuchus cupido cupido),* the nominate East Coast race of Greater Prairie-Chicken, the last of which was seen on Martha's Vineyard in 1932. A subspecies that became extinct more recently was the distinctive "Dusky Seaside Sparrow" *(Ammodramus maritimus nigrescens),* which had a small range on Florida's Atlantic coast. Various mismanaged water policies led to a dramatic and sudden decline of this subspecies; the last individual died in captivity in 1987.

Magnolia Warbler is a monotypic species—with no described subspecies—and thus is only known by the scientific binomial *Dendroica magnolia.* Hardly partial to magnolias, both its English and scientific names derive from Alexander Wilson, who first described it to science in 1811 after having collected one in a magnolia tree in Mississippi. The colorful individual shown here is an alternate-plumaged male. *(Maine, June)*

Author Citations Many distributional and other ornithological works will often include a person's name after the common and scientific names are given. This is the individual (or individuals) who named the species or the subspecies. The name is often followed by the year of publication of the name. The seventh edition of the AOU's *Check-list to North American Birds* (1998) includes the author(s) along with the reference for each genus and species. The fifth edition of the AOU's *Check-list* (1957) includes the citations for the subspecies as well.

Species Concepts Defining what is and what isn't a species has long been a matter of intense debate. Two popular yet very different species concepts have developed in recent years.

■ **Biological Species Concept (BSC)** This species concept defines a species as a genetically cohesive group of populations that are reproductively isolated from other such groups. With the Biological Species Concept, geographic isolation leads to genetic change; if and when these different groups later come back into contact, reproductive isolating mechanisms such as vocalizations, timing of migration, and so on will serve to maintain the essential integrity of the separate groups of populations as distinct species. This species concept allows for some interbreeding, sometimes extensive interbreeding, and these hybrids can indeed be fertile; but within a contact zone, there are enough pure birds of either species to indicate that isolating mechanisms are taking place. The AOU's Committee on Classification and Nomenclature has unanimously endorsed the BSC.

■ **Phylogenetic Species Concept (PSC)** In 1983, ornithologist Joel Cracraft proposed the Phylogenetic Species Concept, which offers a more narrow definition of *species* as "the smallest diagnosable cluster of individual organisms within which there is a parental pattern of ancestry and descent." Suffice it to say, this species concept would recognize many currently recognized subspecies as full species.

Regardless of one's species concept view, ornithologists are increasingly using various lines of genetic evidence as tools in determining species limits and in building classifications that more accurately reflect the evolutionary history of the component species. If a new proposed treatment is based on *published* (and peer-reviewed) evidence, the AOU's Committee on Classification and Nomenclature will evaluate and judge it; the committee's decisions are published in the annual supplements. Its most recent supplement, the 47th (2006), split the species formerly known as Blue Grouse into two species, the more coastal Sooty Grouse *(see page 209)* and the interior Dusky Grouse.

Seaside Sparrows exhibit striking geographical variation. The nominate, ***maritimus*** (bottom), is one of the dull Atlantic races. "Dusky Seaside," (*nigrescens*, top) was dark and heavily streaked. Sadly, "Dusky Seaside," from the Merritt Island region of coastal Florida, went extinct in June 1987. (top: Florida, month unknown; bottom: Connecticut, June)

Higher-Level Classification In addition to effecting changes at the species level, genetic evidence has been of particular value in determining higher-level classifications—those above the level of genus and species, like the composition and relationships of avian families and orders. For instance, on the basis of recent genetic studies, jaegers (including our skuas) are found to be more closely related to auks, murres, and puffins (family Alcidae) than to the gulls and terns of the family Laridae—despite jaegers' gull-like appearance. Because the AOU's

Committee on Classification and Nomenclature is a conservative body, recognizing the importance of stability in names and taxonomic rankings, it requires the publication of at least two independent lines of evidence before any higher-level systematic changes are made.

International Code of Zoological Nomenclature All responsible taxonomists assiduously follow the *International Code of Zoological Nomenclature* (third edition, 1985) and the decisions of the International Commission on Zoological Nomenclature as published in the *Bulletin of Zoological Nomenclature*. Among the more interesting aspects of this code is the Law of Priority, whereby the date of publication of a scientific name determines its validity, with the earliest publication having priority. This hasn't been much of a problem in recent years, in which new taxa are much more infrequently named, but was a big problem in the heyday of species descriptions. For instance, both John James Audubon and John Kirk Townsend independently named a new species, now known as MacGillivray's Warbler. Audubon named it after a Scottish friend, ornithologist William MacGillivray, but Townsend named it for fellow naturalist and medical doctor William Tolmie. Townsend's publication predated Audubon's, and because of the rules of nomenclatural priority, Townsend's scientific name prevailed: *Oporornis tolmiei*. Audubon's tribute to MacGillivray is preserved in the species' English name. Nomenclatural priority also determines the names that prevail when species or genera are "lumped." For example, when the AOU recently determined that Blue Grosbeak (formerly *Guiraca caerulea*) and the buntings in the genus *Passerina* were best treated in the same genus, the older generic name—that having nomenclatural priority—was *Passerina*. Thus, Blue Grosbeak became *Passerina caerulea*. The above decision is also representative of the fact that while the overall number of species has increased worldwide in recent decades, almost entirely as a result of new splits, the number of genera has been reduced.

Common and English Names The *common* name is one or more names that are given to a species apart from the scientific name. For species that occur widely, there may be many common names, largely reflecting that they occur in different countries where different languages are spoken. All of the world's birds have been given an *English* name, thus *Turdus migratorius* is known as American Robin. While we term these English names, the names themselves don't have to be actually in English: Recently, the English names for Hawaiian honeycreepers in the subfamily Drepanidinae were replaced by Hawaiian names that are totally foreign to English speakers. There can be multiple English names for a species, too. Thus the monotypic species we know as Common Loon is known in the United Kingdom as Great Northern Diver. The AOU's Committee on Classification and Nomenclature, responsible for establishing officially used English names, is not bound by the code. It could change these names every other week, if it so chose, but this would be unwise. Perhaps no decision was more unpopular in recent years than the change of the popular English name of Oldsquaw to Long-tailed Duck, the English name used in the United Kingdom. Of course, various philosophies do prevail from time to time. In the 1960s the committee adopted a policy of hostility to patronyms, species names commemorating individuals. Many of these English names were changed in favor of more descriptive (of plumage), yet more bland, names. Some members of the committee hope to restore the former names, which had long been in use.

Recently, with the publication of *Birds of the World: Recommended English Names* (2006) on behalf of the International Ornithological Congress, there has been an effort to establish a single English name worldwide for each species. Sometimes the name adopted is a blend of the names being used in North America and the United Kingdom; thus our Common Loon (Great Northern Diver in the United Kingdom) becomes the Great Northern Loon. In view of the uproar caused by the English name change of Oldsquaw, we don't expect the Committee on Classification and Nomenclature to act quickly in adopting this set of names. Anyone confused by multiple English names should always check the scientific name, adopted for just this situation. The committee has changed English names for species that are mainly peripheral to North America—those that have well-established English names elsewhere within the core of their range. Thus the Old World Rufous-necked Sandpiper, known mainly in North America only as an uncommon visitor to western Alaska, was changed first to Rufous-necked Stint and finally to Red-necked Stint, in concordance with the Old World English name. No doubt the debate on this subject will continue long into the future.

Scientific Collections

Various institutions across North America and Europe and throughout the world house large numbers of scientific specimens of birds. These specimens are usually prepared as *study skins* (as opposed to *life mounts,* which are in the public displays), but they may also be other preparation types, such as complete or partial skeletons or fluid-preserved whole specimens; they are protected from light, insect infestations, and other damaging agents and stored in drawers or in cases. If properly curated and conserved, they should be useful for scientific study for hundreds of years. Material housed in these institutions is the foundation for any serious researcher interested in any of a wide variety of subjects, whether it be taxonomy, evolution, conservation biology, or field identification. Valuable specimens of rare or even extinct species are preserved for future study, and collections made many decades earlier (even 150 years) provide an invaluable record of the birdlife of areas that have been severely altered by humans. Furthermore, most institutions are now preserving tissue samples, which are vital to the burgeoning field of research on avian genetics and molecular taxonomy.

All five editions of the *National Geographic Field Guide to the Birds of North America* relied primarily on museum specimens for reference material for the rendering of illustrations by artists. Many institutions provided specimens, but foremost among them was the National Museum of Natural History in Washington, D.C., followed by the Louisiana State University Museum of Zoology in Baton Rouge, and the Field Museum of Natural History in Chicago. All of these institutions (including those not named here) deserve our support and thanks.

This light-morph adult **Parasitic Jaeger** is now placed in the skuas and jaegers family (Stercorariidae). Formerly, skuas and jaegers were considered to be a subfamily of the gulls, terns, and skimmers (Laridae). Recent genetic studies suggest that skuas and jaegers are more closely related to the Alcidae family of auks, murres, and puffins. They were assigned to their own family in 2006. *(Alaska, June)*

ABA American Birding Association; publishes *Birding*.

accidental An exceptional occurrence. In the ABA Checklist Area, a species that has been recorded five or fewer times overall, *or* fewer than three times in the past thirty years.

alternate plumage Plumage worn by an adult bird during the breeding season, produced by a partial molt before breeding. Breeding and nuptial plumage are synonyms.

AOU American Ornithologists' Union; publishes the *Check-list of North American Birds,* now in its seventh edition.

apical spot A spot at the tip of a feather.

auriculars A group of lacy feathers covering the ear (aural) openings; often bordered with contrasting stripes or lines. Ear coverts and cheek are synonyms.

axillaries A cluster of feathers in the bird's "armpit"; they are recognizably longer than the underwing coverts.

bare parts Those areas of a bird's body completely without feathers.

basic plumage For most birds, the plumage worn for the longest time each year and produced by a complete molt. Nonbreeding and winter plumage are synonyms.

bend of the wing The joint between the outer wing and the inner wing, where a bird's wing angles back noticeably. Wrist and carpal are synonyms.

binomial nomenclature The scientific name for an organism consisting of two words—the first name being the name of the genus, the second being the specific epithet for species name.

biological species concept (BSC) Defines a species as a genetically cohesive group of populations that is *reproductively isolated* from other such groups. See *phylogenetic species concept.*

breeding plumage See *alternate plumage*, a synonym.

call notes Bird sounds that are generally shorter than songs and seem to convey a specific message, such as begging calls, alarm calls, and contact calls (or chip notes); these are generally innate rather than learned.

carpal bar A bar on the inner wing formed by contrasting secondary coverts.

casual Not annual, but records reflect a pattern of occurrence. In the ABA Checklist Area, six or more records that *include* three or more records in the past 30 years.

cere A band of skin covering the base of the bill.

cline A gradual change in certain characteristics of individuals of the same species, which is evident in a geographic progression from one population to the next.

common name The names other than the scientific names by which an organism is called; these include the standard English names adopted by the AOU.

coverts The small feathers that partially overlay the flight feathers of the wing and tail at their bases.

culmen Ridge of the upper mandible from base to tip.

dimorphic Two forms of the same species or population that differ in some aspect of plumage, size, or shape. Sexually dimorphic species show fixed differences between the sexes. Compare to *polymorphic*.

ear coverts See *auriculars*, a synonym.

endemic Restricted to a given geographic area.

eye crescent A partial eye ring visible either above or below—or both above and below—the eye.

eye line A colored line that passes through the eye.

eye ring A circle of colored feathers encircling the eye.

family Level of classification above genus and below order, in which evolutionarily related genera are placed.

flight feathers The major feathers of the wing (primaries, secondaries) and the tail feathers (or rectrices).

flight note Calls usually given by a bird in flight, whether during its nocturnal migration or from perch to perch.

gape The juncture of the maxilla and the mandible.

genus (plural, *genera*) The level of classification above the species level and below the family level.

gonys The ridge formed where the two segments of the lower mandible join.

greater coverts A single row of feathers that partially overlays the flight feathers of the wing.

humerals A third set of flight feathers on the uppermost wing bone (humerus) that overlaps the secondaries as the wing folds; well developed in long-winged birds.

Humphrey-Parkes System System for naming plumages and molts that does not use terms related to seasons. See *basic plumage* and *alternate plumage*.

hybrid The result of breeding between individuals of different species. First-generation hybrids are known as F1 hybrids.

immature Not fully adult, either in some specific area of development, such as plumage, or as a whole.

iris The colored area of the eye surrounding the pupil.

juvenal plumage Feather coat worn by juvenile birds after they have molted their natal down; it consists of the first true contour feathers.

lateral crown stripe A stripe of darker feathers along the sides of the crown and above the supercilium.

lesser coverts Multiple rows of overlapping feathers of the wing that partially overlay the median coverts.

leucism An abnormal paleness occurring in a bird's plumage due to a dilution of pigmentation.

lores Small area located between the eye and the base of the bill.

malar stripe A distinctively colored stripe located below the submoustachial stripe and adjacent to the throat.

mandible The lower half of the bill.

mantle An alternate term often used to describe the back, scapulars, and upperwing coverts of gulls as a whole.

maxilla The upper part of the bill, also called the upper mandible.

median coverts A single row of feathers lying between the lesser and greater coverts of the wing.

median crown stripe A contrasting line of feathers down the center of the crown.

mirror The subterminal white spots near the primary tips of many gull species.

molt The orderly replacement of old feathers with new feathers.

monotypic A taxonomic category that contains only a single taxon; a monotypic species has no subspecies.

morph Usually describes a fixed color variation within a population or entire species.

moustachial stripe A dark facial stripe that extends from the gape and follows the lower border of the auriculars.

nail The hard, hooked tip of the upper mandible.

nominate subspecies First type of the species to be described for a polytypic species; the subspecies name is the same as the second word of the species name.

nonbreeding plumage See *basic plumage*, a synonym.

orbital ring Ring of bare skin immediately surrounding the eye. Compare to *eye ring*.

passerine All birds in the order Passeriformes, known as perching birds, which includes the songbirds.

patagium The leading edge of the inner wing; if this area is conspicuously dark it forms a patagial bar.

pelagic Of the ocean.

phylogenetic species concept (PSC) Defines a species as "the smallest diagnosable cluster of individual organisms within which there is a parental pattern of ancestry and descent." Compare to *biological species concept*.

plumage The collective term for all the feathers that cover a bird's body; also known as pterylosis.

polymorphic Species or populations showing fixed variations, such as color morphs.

polytypic A taxonomic category containing two or more representatives of the category immediately below it. A polytypic species contains two or more subspecies.

postocular stripe A stripe extending back from the eye.

prealternate molt An annual partial molt in the late winter and early spring that results in alternate plumage. Not all species have a prealternate molt.

prebasic molt An annual complete molt, usually in late summer or early fall, that results in basic plumage.

primaries The long flight feathers of the outer wing.

primary coverts Small feathers arranged in rows that overlay the base of the primaries.

primary projection The length of the primaries projecting beyond the longest tertial feather on the folded wing.

rare A category denoting abundance scarcer than uncommon but more numerous than casual. Refers to species that occur annually but in low numbers.

rectrices (singular, *rectrix*) The long flight feathers of the tail; tail feathers is a synonym.

remiges (singular, *remix*) The flight feathers of the wing—primaries and secondaries.

rump The area between the back and the uppertail coverts.

scapulars A group of feathers that overlays the area where the wing attaches to the body.

scientific name The two-part species name devised by Linnaeus. See binomial nomenclature.

secondaries The flight feathers of the inner wing.

song Patterned vocalizations usually given by males to attract mates or defend a territory.

species Level of classification below genus and above subspecies. See *biological species concept* and *phylogenetic species concept* for differing definitions of species.

spectacles A pale eye ring that connects with the supraloral area, contrasting with the rest of the face.

subadult A bird that has not attained full adult plumage.

submoustachial stripe A stripe (usually pale) located below the moustachial stripe and above the malar stripe.

subspecies The taxonomic level below species, containing individuals from a particular geographic region where the large majority of individuals are morphologically distinct from other individuals of the same species from a different area. *Race* is a synonym.

supercilium Pale feathers forming a stripe on the side of the head above the eye. Eyebrow is a synonym.

supraloral The part of the supercilium between the eye and the bill; located just above the lores.

tarsus (plural, *tarsi*) The section of leg directly above a bird's foot (the upper section of the avian foot).

taxonomy The classification of organisms—assigning names and relationships.

tertials Prominent feathers, usually three in number, that overlay the secondaries on the folded wing; often referred to as the innermost secondaries.

tibia The section of leg above the tarsus.

trinomial The three-word scientific name, the third word of which designates the organism's subspecies.

undertail coverts The covert feathers that cover the bases of the tail feathers from below. Crissum is a synonym.

underwing coverts The covert feathers on the underside of the wing that cover the bases of the primaries and secondaries. Wing linings is a synonym.

uppertail coverts The covert feathers that cover the bases of the tail feathers from above; located between the rump and the tail.

vent The region located where the belly feathers meet the undertail coverts.

wing bars One or two contrasting bars running across a bird's wing. The effect results from the pale tips on the greater and/or median secondary coverts.

Many of the glossary definitions have been excerpted from the Handbook of Bird Biology, from the Cornell Lab of Ornithology.

ADDITIONAL READING

Below is an abbreviated list of books that interested readers might wish to consult; most specialized books and guides are omitted. Web sites will keep you up to date with current sightings and rare-bird information. Here we list a few of the main Internet portals.

Alderfer, Jonathan, ed. *Complete Birds of North America*. National Geographic Society, 2005.

American Birding Association [ABA]. *ABA Checklist: Birds of the Continental United States and Canada*, 6th edition. ABA, 2002.

American Ornithologists' Union [AOU]. *Check-list of North American Birds*, 7th edition. AOU, 1998. Standard taxonomic authority for North American birds. The fifth edition (1957) is the last edition with a complete list of subspecies.

Beadle, David, and James Rising. *Sparrows of the United States and Canada: The Photographic Guide*. Academic Press, 2002.

Brinkley, Edward S. *Field Guide to the Birds of North America*. Sterling Publishing, 2007. A photographic field guide.

Campbell, Bruce, and Elizabeth Lack. *A Dictionary of Birds*. Buteo Books, 1985.

del Hoyo, Josep, Andrew Elliot, Jordi Sargatal, and David Christie, eds. *Handbook of the Birds of the World*. Volumes 1–11. Lynx Ediciones, 1992–2006. Comprehensive treatment of all bird species, five additional volumes planned; available online at www.hbw.com/index.html.

Dickinson, Edward C. *The Howard and Moore Complete Checklist of the Birds of the World*, 3rd ed. Princeton University Press, 2003. Checklist with subspecies and ranges.

Dunn, Jon L., and Jonathan Alderfer, eds. *Field Guide to the Birds of North America*, 5th edition. National Geographic Society, 2006.

Dunn, Jon L., and Kimball L. Garrett. *A Field Guide to Warblers of North America*. Houghton Mifflin, 1997.

Dunne, Pete. *Pete Dunne on Bird Watching*. Houghton Mifflin, 2003.

Dunne, Pete, David Sibley, and Clay Sutton. *Hawks in Flight*. Houghton Mifflin, 1988.

Garrett, Kimball, and Jon Dunn. *Birds of Southern California: Status and Distribution*. Los Angeles Audubon Society, 1981.

Harrison, Peter. *Seabirds: An Identification Guide*. Houghton Mifflin, 1983.

Hayman, Peter, John Marchant, and Tony Prater. *Shorebirds: An Identification Guide to the Waders of the World*. Croom Helm, 1986.

Hoffmann, Ralph. *Birds of the Pacific States*. Houghton Mifflin, 1927.

Howell, Steve N. G. *Hummingbirds of North America: The Photographic Guide*. Academic Press, 2002.

Howell, Steve N. G., and Jon Dunn. *Gulls of the Americas*. Houghton Mifflin, 2007.

Kroodsma, Donald. *The Singing Life of Birds*. Houghton Mifflin, 2005.

Leahy, Christopher W. *The Birdwatcher's Companion to North American Birdlife*. Princeton University Press, 2004.

Ligouri, Jerry. *Hawks from Every Angle*. Princeton University Press, 2005.

Mullarney, Killian, Lars Svensson, Dan Zetterström, and Peter J. Grant. *The Complete Guide to the Birds of Europe*. Princeton University Press, 2000. Considered by many to be the best field guide ever written.

O'Brien, Michael, Richard Crossley, and Kevin Karlson. *The Shorebird Guide*. Houghton Mifflin, 2006.

Poole, Alan, and Frank Gill, eds. *The Birds of North America*. The Academy of Natural Sciences and the American Ornithologists' Union, 1992–2002. Accounts of all birds breeding in North America, including Hawaii. Published individually by species; available online at bna.birds.cornell.edu/BNA.

Pough, Richard H. *Audubon Bird Guide: Eastern Land Birds*. Doubleday, 1946. First in a series of classic but out-of-print bird guides with illustrations by Don Eckelberry.

Pyle, Peter. *Identification Guide to North American Birds, Part I*. Slate Creek Press, 1997. In-depth banders' guide to North American passerines and near-passerines; extensive treatment of subspecies.

Robbins, Chandler S., Bertel Bruun, and Herbert S. Zim. *Birds of North America*. Golden Press, 1966. The only North American field guide to include sonograms.

Sibley, David A. *The Sibley Guide to Birds*. Alfred A. Knopf, 2000.

Sibley, David A. *Sibley's Birding Basics*. Alfred A. Knopf, 2002.

Williamson, Sheri L. *Hummingbirds of North America*. Houghton Mifflin, 2001.

Web Sites

American Birding Association.
www.americanbirding.org

American Ornithologists' Union.
www.aou.org/checklist/index.php3

Birding on the Net.
www.birdingonthe.net/index.html

Breeding Bird Survey. www.pwrc.usgs.gov/BBS

Cornell Laboratory of Ornithology.
www.birds.cornell.edu

eBird. www.ebird.org

Frontiers of Identification.
www.birdingonthe.net/mailinglists/FRID.html

SORA (Searchable Ornithological Research Archive). elibrary.unm.edu/sora/index.php

INDEX

Boldface indicates illustrations.

We dedicate this book to Kimball L. Garrett,
friend, colleague, and birding companion for decades

Acknowledgments

We wish to thank the following individuals whose help was essential to us. Kimball L. Garrett of the Los Angeles County Museum of Natural History made many insightful suggestions, both before and after the text was drafted, and read through the entire text twice. Bob Steele of Inyokern, California, our photographic consultant and an expert bird photographer, was tireless in his pursuit of the best available bird photographs. Zora Margolis Alderfer supplied editorial expertise as the book was written and offered many constructive comments from the perspective of a beginning birder. Russell Galen of New York City, New York, our literary agent, was enthusiastic and exacting in getting this project started.

We would also like to thank the following individuals who gave freely of their time and good advice. Tom and Jo Heindel of Big Pine, California, read and commented on chapter 3 and offered many valuable suggestions. Stephen Ingraham of Zeiss Optics reviewed our material on birding optics and supplied a helpful binoculars image. Louis Bevier of Fairfield, Maine, tracked down a superb photograph of the recent New Hampshire Western Reef-Heron taken by Lillian Stokes. Kevin Karlson of Rio Grande, New Jersey, photographed a number of images specifically for this book. The New Jersey Audubon Center for Research and Education loaned binoculars and scopes for photographs, and its employees Lillian Armstrong, Beth Ciuzio, Brian Moscatello, and Dale Rosselet graciously modeled them.

Illustration Credits

Abbreviations for terms appearing below: *t*-top; *b*-bottom; *l*-left; *r*-right; *c*-center

Jonathan Alderfer: 27, 51, 66*b*, 85*b*, 198–199; **Juan Bahamon:** 49; **James T. Blair**/NGS Image Collection: 7; **Rick and Nora Bowers**/VIREO: 101; **Richard Crossley:** 9, 39, 67*t*, 68*l*, 74, 87, 113, 114, 135, 169*l*, 173, 190, 195; **Rob Curtis**/The Early Birder: 29*l*; **Mike Danzenbaker:** 38, 52, 68*r*, 78, 83, 105*r*, 161*l*, 165*l*, 165*r*, 170; **Richard Day**/Daybreak Imagery: 17; **Robert H. Day:** 66; **Christopher Dodds:** 140; **Jon L. Dunn:** 206; **Christina L. Evans:** 62; **Heather Forcier:** 148, 181; **Paul J. Fusco:** 2–3, 12, 15, 64, 72, 120*t*, 138, 151*tr*, 186*t*, 212*b*; **Greg Gillson**/The Bird Guide: 184; **Dr. Hiroshi Hasegawa:** 67*b*; **David Hemmings:** front cover, 159, 207; **Kevin T. Karlson:** 20, 21, 22, 25, 182, 194, 204, 205; **Peter LaTourette**/VIREO: 100; **Greg Lavaty:** 71, 117, 157; **Robert Mansell**/Time Life Pictures/Getty Images: 208; **Robert McCaw:** 6, 56, 116*l*; **Garth McElroy:** 32, 77, 93*t*, 93*b*, 98*l*, 104, 144*l*, 145*t*, 160*r*, 167*t*, 186*b*; **Jody Melanson:** 63*t*, 110; **Alan Murphy:** 4, 8, 47, 63*b*, 92*tl*, 92*tr*, 92*bl*, 92*br*, 94*b*, 102*b*, 105*l*, 108*b*, 130*b*, 137*l*, 137*r*, 139*l*, 141*t*, 167*b*, 168*l*, 183*b*; **Jim Neiger:** 31; **Wayne Nicholas:** 11; **James Ownby:** 155; **E. J. Peiker:** 70*l*, 70*r*, 162*t*, 162*b*; **Powdermill Avian Research Center:** 153*b*, 197; **Robert Royse**; 28*l*, 28*r*, 35*r*, 40, 41, 42*l*, 42*r*, 46, 48, 50, 53, 58, 61, 65*l*, 65*r*, 75, 76, 79*t*, 79*b*, 80, 81, 82, 86, 89*t*, 89*b*, 97*t*, 97*b*, 99*r*, 102*t*, 102*c*, 106*t*, 106*c*, 106*b*, 108*c*, 108*t*, 118, 119*l*, 119*r*, 123, 124, 139*r*, 141*b*, 151*tl*, 168*r*; **John A. Ruthven:** 142; **Larry Sansone:** 54*t*, 54*b*, 147*t*, 147*b*, 160*l*, 193; **Brian E. Small:** 10, 13, 57*l*, 111*t*, 111*b*, 116*r*, 125*l*, 125*r*, 128*l*, 128*r*, 134*l*, 134*r*, 143*b*, 144*r*, 163*t*, 183*t*, 185, 211; **Bob Steele:** 16, 29*r*, 30, 33, 44, 45, 57*r*, 84, 85*t*, 91, 95*t*, 98*r*, 99*l*, 115, 127, 133, 136, 143*t*, 145*b*, 150, 152, 153*t*, 154, 161*r*, 163*b*, 169*r*, 177, 179, 209; **Lillian Stokes:** 196; **P. W. Sykes, Jr.**/VIREO: 212*t*; **Daniella Theoret:** 69; **Tulsa Audubon Society:** 200; **Tom Vezo:** 171; **Brian K. Wheeler**/VIREO: 95*b*; **Matthew K. Whitley, M.D.:** 35*l*; **Zeiss Optics:** 19; **Jim Zipp:** 73*t*, 73*c*, 73*b*, 94*t*, 120*b*, 130*t*, 151*bl*, 151*br*, 175, 215.

Additional Credits

page 25: Title page from *Birds of the Pacific States* by Ralph Hoffmann, Houghton Mifflin, 1927. Painting by Allan Brooks.

page 37: Chart adapted from *Birds of Southern California: Status and Distribution*, by Kimball Garrett and Jon Dunn, Los Angeles Audubon Society, 1981.

page 142: Frontispiece illustration from *Life Histories of North American Cardinals, Grosbeaks, Buntings, Towhees, Finches, Sparrows, and Allies*, by Arthur Cleveland Bent, Dover Publications, 1968. Painting by John A. Ruthven.

page 146: Data for the breeding range map of Fox Sparrow subspecies from "Species Limits in the Fox Sparrow" by R. M. Zink and A. E. Kessen, in *Birding* 31:508-517 (1999); data for winter ranges from "Call Notes and Winter Distribution in the Fox Sparrow Complex" by Kimball L. Garrett, Jon L. Dunn, and Robert Richter, in *Birding* 32:412–417 (2000).

p. 189: Data for the migration map, eastern section, courtesy of Jon L. Dunn and Kimball L. Garrett, *A Field Guide to the Warblers of North America*, Houghton Mifflin Company, Boston, 1997.

National Geographic Birding Essentials
Jonathan Alderfer and Jon L. Dunn

Published by the National Geographic Society
John M. Fahey, Jr., *President and
    Chief Executive Officer*
Gilbert M. Grosvenor, *Chairman of the Board*
Nina D. Hoffman, *Executive Vice President;
    President, Book Publishing Group*

Prepared by the Book Division
Kevin Mulroy, *Senior Vice President and Publisher*
Leah Bendavid-Val, *Director of Photography
    Publishing and Illustrations*
Marianne R. Koszorus, *Director of Design*

Barbara Brownell Grogan, *Executive Editor*
Elizabeth Newhouse, *Director of
    Travel Publishing*
Carl Mehler, *Director of Maps*

Staff for This Book
Barbara Levitt, *Editor*
Jennifer Conrad Seidel, *Text Editor*
Carol Norton, *Birding Program Art Director*
Jennifer Davis, Bob Steele, *Illustrations Editors*
Richard S. Wain, *Production Project Manager*
Rob Waymouth, *Illustrations Specialist*
Nicole DiPatrizio, *Design Assistant*

Jennifer A. Thornton, *Managing Editor*
Gary Colbert, *Production Director*

Manufacturing and Quality Management
Christopher A. Liedel, *Chief Financial Officer*
Phillip L. Schlosser, *Vice President*
John T. Dunn, *Technical Director*
Chris Brown, *Director*
Maryclare Tracy, *Manager*
Nicole Elliott, *Manager*

Special Contributors
Kimball L. Garrett, Barbara Seeber *(text)*;
J. Naomi Linzer *(index)*

Founded in 1888, the National Geographic
Society is one of the largest nonprofit scientific
and educational organizations in the world.
It reaches more than 285 million people worldwide each month through its official journal,
NATIONAL GEOGRAPHIC, and its four other
magazines; the National Geographic Channel;
television documentaries; radio programs; films;
books; videos and DVDs; maps; and interactive
media. National Geographic has funded more
than 8,000 scientific research projects and
supports an education program combating
geographic illiteracy.

For more information, please call
1-800-NGS LINE (647-5463)
or write to the following address:

National Geographic Society
1145 17th Street N.W.
Washington, D.C. 20036-4688 U.S.A.

Visit us online at www.nationalgeographic.com/books

For information about special discounts for
bulk purchases, please contact
National Geographic Books Special Sales:
ngspecsales@ngs.org

For rights or permissions inquiries, please contact
National Geographic Books Subsidiary Rights:
ngbookrights@ngs.org

Copyright © 2007 National Geographic Society.
All rights reserved. Reproduction of the whole or any
part of the contents without written permission from
the publisher is prohibited.

Library of Congress
Cataloging-in-Publication Data
Alderfer, Jonathan K.
    National Geographic birding essentials/Jonathan Alderfer
    and Jon L. Dunn.
        p. cm.
    Includes bibliographical references and index.
    ISBN 978-1-4262-0135-6
1. Bird watching. 2. Birds--Identification. I. Dunn, Jon,
1954- II. National Geographic Society (U.S.) III. Title.
IV. Title: Birding essentials.
QL677.5.A395 2007
598.072'34--dc22
                                    2007030960

Printed in U.S.A.